WITHDRAWN
FROM THE LIBRARY OF
UNIVERSITY OF ULSTER

Ion Exchange for Pollution Control

Volume II

Editors

Calvin Calmon

Environmental Consultant
Princeton, New Jersey
and
Partner, Water Purification Associates
Cambridge, Massachusetts

Harris Gold

Partner, Water Purification Associates
Cambridge, Massachusetts

Editor-in-Chief
Water Pollution Control Technology Series

Richard Prober

Adjunct Professor
Department of Chemical Engineering
Case Western Reserve University
Cleveland, Ohio

CRC PRESS, INC.
Boca Raton, Florida 33431

Library of Congress Cataloging in Publication Data

Main entry under title:

Ion exchange for pollution control.

Bibliography: p.
Includes index.
1. Sewage--Purification--Ion exchange process.
I. Calmon, Calvin. II. Gold, Harris. III. Series.
TD757.5.155 628'.3 78-13258
ISBN 0-8493-5153-7 (v. 1)
ISBN 0-8493-5154-5 (v. 2)

This book represents information obtained from authentic and highly regarded sources. Reprinted material is quoted with permission, and sources are indicated. A wide variety of references are listed. Every reasonable effort has been made to give reliable data and information, but the author and the publisher cannot assume responsibility for the validity of all materials or for the consequences of their use.

International Standard Book Number 0-8493-5153-7 (Volume I)
International Standard Book Number 0-8493-5154-5 (Volume II)

Library of Congress Card Number 78-13258
Printed in the United States

FOREWORD

These volumes in the Uniscience series on Water Pollution Control Technology are a comprehensive report on applications of ion exchange to treatment of wastewater. They cover existing applications, and by extension of fundamental principles, they show the basis for innovative technology. The authors have been principals in the development of ion-exchange materials and process technology.

Our objective in these volumes and in subsequent volumes of the series is to provide a reference manual for design engineers, planners, and managers in industry and government. This is particularly important in the present critical period for implementation of water pollution control.

Richard Prober
Cleveland, Ohio
July 21, 1978

PREFACE

There are many books on ion exchange dealing with general subject matter and special topics such as the role of ion exchange in analytical, organic, and biochemical applications. While some aspects of ion exchange in pollution control were discussed in these texts, especially with reference to the removal of toxic heavy metals, the purification of metal plating baths, and the recovery of metal ions from metal plating rinse waters, the subject as a whole was never covered. In view of the emphasis in recent years on pollution control, the application of ion exchange in this endeavor has grown so a book on the subject itself is timely.

The aim of these volumes is not to cover all phases of ion-exchange theory, which may be found in in general texts, nor to cover every application in the literature, or to show an engineer ways on how to become an expert in the field so he could "do it all by himself". The main purpose of these books is to show the practical engineer what has been done in various types of applications of ion-exchange processes in pollution control, how to set up laboratory tests, the problems that may be encountered, to identify the individuals and organizations who are experts in the various phases of ion exchange, and most importantly, to emphasize the new developments in polymers with active sites that offer new approaches to wastewater treatment methods.

These books contain over forty chapters written by researchers and engineers from the industrial, academic, and research communities in the United States, United Kingdom, Australia, Canada, and Poland. The books are divided into six major sections:

General Aspects — Twelve chapters dealing with the ion-exchange process, ion exchangers, equipment, systems, and economic factors.

Inorganic Applications — Seven chapters dealing with the removal of heavy metals and ammonia and nitrate recovery from a variety of wastewaters.

Organic Applications — Contains four chapters on base recovery from spent sulfite liquors, decolorizing wood pulp bleaching effluents, and the use of polymeric adsorbents.

Other Applications — Seven chapters involving the treatment of effluents from nuclear and municipal treatment plants, agricultural wastewaters and the use of ion exchange for laboratory and analytical evaluations.

Uses of Polymers with Active Sites — Three chapters covering coagulants, dewatering agents, and dispersants.

New Developments — Ten chapters dealing with novel exchangers and processes developed in the last decade.

In addition, the volumes contain a bibliography of ion-exchange text books, a listing of ion exchanger and equipment manufacturers, and special tables of data. We have broadened the aspect of the subject to include polymeric polyelectrolytes as we feel it is an extension of ion exchange, and future developements in this field for wastewater treatment will increase with time.

The authors chosen to write the chapters in these volumes are well known for their contributions in the application of polymers with active sites to pollution control. Credit must be given to both the authors and the organizations they represent. The authors, in many instances, wrote their chapters on their own time.

We hope that these volumes will make a contribution to improving the quality of our environment and to saving our resources which grow scarcer every day.

Calvin Calmon
Harris Gold
Cambridge, Massachusetts
July, 1978

EDITOR-IN-CHIEF

Richard Prober is a principal engineer with GMP Associates, and Adjunct Professor of Chemical Engineering at Case Western Reserve University, Cleveland, Ohio.

Dr. Prober received his B.S. in chemical engineering in 1957 from the Illinois Institute of Technology. In 1958 he received his M.S. degree and in 1962 his Ph.D. degree in chemical engineering from the University of Wisconsin.

Dr. Prober's accomplishments include curriculum development for wastewater treatment plant operator training; development of low-flow dissolved oxygen models for the Cuyahoga River and Tinkers Creek, including direction of stream surveys to calibrate the models; development of process-design oriented B.S. and graduate level programs in wastewater engineering; and extensive research into process development of activated carbon treatment and treatment for industrial wastes containig cyanides.

His professional associations include the Water Pollution Control Federation, American Institute of Chemical Engineers, and the American Chemical Society. Dr. Prober has also served as Symposium Chairman for national meetings of the U.S. Environmental Protection Agency, American Institute of Chemical Engineers, and Wastewater Equipment Manufacturer's Association.

THE EDITORS

Calvin Calmon, Ph.D., has been in the field of industrial chemical research since he received his Ph.D. degree in physical chemistry from Yale University in 1938. (B.A. from Dartmouth College in 1934.) Except for a period of service from 1944 to 1947 as an officer in the U.S. Army, until his retirement in 1973, he worked continuously with various divisions of the Sybron Corporation. Until 1960 he was with the Permutit Company, where he was Head of Laboratories. He then joined Ionac Chemical Company as Director of Research, later became Vice President of Research and, on retirement, was Senior Vice President and Research Consultant to the Chemical Group of Sybron Corporation. In addition to this various consulting activities he is a limited partner in Water Purification Associates.

Most of his industral work, including about 65 technical publications and 19 patents, has been in water treatment, absorbents, ion exchangers, polymers, membmnes, and pollution control. In 1972 he received the Distinguished Service Award from the Environmental Chemistry Section of the American Chemical Society. In 1976 he received from the U.S. Environmental Protection Agency the Certificate of Appreciation for contributions towards the improvement of the environment.

He has served as Chief Research Scientist on contract projects for the Atomic Energy Commission, the Office of Saline Water, the National Aeronautics and Space Administration, and the U.S. Public Health Service. From 1962 to 1972 he served as Section Editor on Water for Chemical Abstracts, published by the American Chemical Society and is currently a scientific and technical merit reviewer for research demonstration grant applications submitted to the U.S. Environmental Protection Agency. With Dr. R. Kressman of England, he edited *Ion Exchangers in Organic and Biochemistry.* During World War II he was Chief of Biochemistry and Serology Sections of the Command Laboratory of the U.S. Army Forces of the Western Pacific. For his chemical work in developing the desalting kit for converting seawater to potable water, now used by all ocean-flying planes, he received a U.S. Government citation. He has received patents for a calcium sensor, improved condensate demineralizing, desalination, pollution control, and ion-exchange processes. He is a member of many technical societies and a Fellow of the American Association for the Advancement of Science.

Harris Gold, Ph.D., is a Partner with Water Purification Associates, Cambridge Massachusetts. He received a B.M.E. from the Polytechnic Institute of New York in 1958, a M.S.M.E. from Columbia University in 1959, and a Ph.D. from the California Institute of Technology in 1963.

From 1963 through 1974, Dr. Gold was with Avco Systems Division, Wilmington, Massachusetts, where he engaged in and directed theoretical and experimental research in heat and mass transfer in fluid flow systems. He was also responsible for the development of a continuous moving bed ion-exchange system for softening pretreatment. Since 1974, Dr. Gold has been with Water Purification Associates. His primary interests include the development of water management systems for power and fuel producing plants and the application of innovative water treatment technologies for industrial wastewater.

Dr. Gold has published many scientific papers and is the co-author of a recent book on the role of water in synthetic fuel production. He is a member of the American Institute of Chemical Engineers, the American Electroplaters Society, and the American Water Works Association.

CONTRIBUTORS

I. M. Abrams, Ph.D.
Manager, Technical Development
Diamond Shamrock Corporation
Functional Polymers Division
Redwood City, California

R. M. Ahlgren, B.S.
Technical Director
Aqua Chem, Inc.
Milwaukee, Wisconsin

J. Barcicki, Ph.D.
Associate Professor
Department of Chemical Technology
Institute of Chemistry
University of Maria Curie-Sklodowska
Lublin, Poland

B. A. Bolto, Ph.D.
Assistant Chief
CSIRO Division of Chemical Technology
South Melbourne, Australia

F. L. Burnett, M.S.
Product Development Manager
Functional Polymers Division
Diamond Shamrock Corporation
Redwood City, California

Randhir C. Chopra, B.Ch.E.
Vice President
Process and Commercial Operations
Chemical Separations Corporation
Oak Ridge, Tennessee

G. J. Crits, M.Ch.E., P.E.
Technical Director
Cochrane Environmental Systems
Crane Company
King of Prussia, Pennsylvania

T. R. Dillman
Vice President
Research and Development
Illinois Water Treatment Company
Rockford, Illinois

E. C. Feeney, M.B.A.
Technical Promotions Manager
Rohm and Haas Company
Philadelphia, Pennsylvania

Sallie A. Fisher, Ph.D.
President
Puricons, Inc.
Berwyn, Pennsylvania

J. W. Frankenfeld, Ph.D.
Senior Research Chemist
Exxon Research and Engineering Company
Linden, New Jersey

J. E. Goddard, B.Sc.
Managing Director
Atlas Process Consultants, Limited
Hounslow, Middlesex, England

I. R. Higgins,B.S.
Vice President and Technical Director
Chemical Separations Corporation
Oak Ridge, Tennessee

A. Himsley
President
Himsley Engineering Limited
Toronto, Ontario

G. Klein, M.S.
Research Engineer
Sea Water Conversion Laboratory
University of California
Richmond, California

R. Kunin, Ph.D.
Consultant
Yardley, Pennsylvania

Norman N. Li, Sc.D.
Senior Research Associate
Exxon Research and Engineering Company
Linden, New Jersey

L. B. Luttinger, Ph.D.
Senior Research Scientist
The Permutit Company, Inc.
Princeton, New Jersey

F. X. McGarvey, M.Ch.E.
Manager
Technical Center
Ionac Chemical Company
Birmingham, New Jersey

W. S. Miller, B.S.
Research Associate
The Permutit Company, Inc.
Research and Development Center
Monmouth Junction, New Jersey

A. B. Mindler, B. Ch. E., P. E.
Research Group Leader
The Permutit Company, Inc.
Research and Development Center
Monmouth Junction, New Jersey

L. Pawłowski, Ph.D.
Assstant Professor
Department of Chemical Technology
Institute of Chemistry
University of Maria Curie-Sklodowska
Lublin, Poland

F. X. Pollio, M.S.
International Sales Manager
Ionac Chemical Company
Birmingham, New Jersey

M. C. Rowe, C.Chem., M.R.I.C.
Marketing Specialist
Portals Water Treatment Ltd.
Isleworth, Middlesex, England

E. Salem, M.Ch.E., M.B.A.
Vice President
Technical Department
Ecodyne Corporation
Graver Water Division
Union, New Jersey

W. Schwoyer, B.S.
Research Group Leader
The Permutit Company, Inc.
Research and Development Center
Monmouth Junction, New Jersey

H. H. Sephton, D.Sc.
Research Chemist and Principal Investigator
Sea Water Conversion Laboratory
University of California
Richmond, California

J. D. Sherman, Sc.D.
Senior Development Consultant
Molecular Sieve Department
Union Carbide Corporation
Tarrytown, New York

G. P. Simon
Manager
Special Applications
Ionac Chemical Company
Birmingham, New Jersey

P. C. Singer, Ph.D.
Professor
Department of Environmental Sciences and Engineering
University of North Carolina
Chapel Hill, North Carolina

G. K. Stephens, Ph.D.
Section Head
Process Development
I.C.I. Australia Limited
Ascot Vale, Victoria, Australia

H. F. Walton, D.Phil.
Professor
Department of Chemistry
University of Colorado
Boulder, Colorado

J. F. Wilkes, M.Ch.E.
Consulting Chemical Engineer
LaGrange, Illinois

R. E. Wing, Ph.D.
Research Chemist
Northern Regional Research Center
U.S. Department of Agriculture
Peoria, Illinois

TABLE OF CONTENTS

VOLUME I

GENERAL ASPECTS

INORGANIC APPLICATIONS

VOLUME II

ORGANIC APPLICATIONS

Organic Applications

Chapter 1

SORPTION OF ORGANIC SUBSTANCES

F. X. Pollio

TABLE OF CONTENTS

I. INTRODUCTION

The use of ion-exchange resins is certainly not limited to conventional water softening and demineralization applications with which most of us are well familiar. Ion-exchange resins are also most useful for reversibly exchanging or sorbing organic matter from liquids. This chapter, accordingly, is devoted to the uptake of organic substances from water and nonaqueous solutions. The removal of inorganic solutes from such organic-bearing systems as glycerine, gelatin, sugar, and similar types of solutions will not be discussed.

Through the use of ion-exchange resins, a liquid may be purified from an organic impurity, and the organic substance may be concentrated so that it can be recovered for a useful purpose or discarded, if need be, into the environment. Ion-exchange resins can also be employed for separating and isolating organic substances whose recovery otherwise, using conventional techniques, might represent a problem. The use of ion-exchange resins for these applications has not been limited to aqueous systems particularly since the introduction of macroporous type ion-exchange resins toward the end of the 1950s.

Prior to the availability of the macroporous ion-exchange resins, which exhibit definite porous properties in media other than water, the utility of gel or conventional type exchangers was usually limited to aqueous or polar liquid systems because the gel or conventional type resins exhibited unfavorable kinetics of sorption for organic species from liquids nonpolar in character. In addition, the resins were found to suffer from physical instability with prolonged cyclic use, particularly where the ion-exchange resins were regenerated with an aqueous or polar regenerant solution.

II. SORPTION FROM AQUEOUS MEDIA

A. Organics in Water Supplies

Most water supplies have at least some amount of organic matter present. The quantity may vary from a few to several hundred parts per million, depending on the geographic location and the type of water source. While well waters may have only trace amounts of organic matter, surface waters may contain up to several hundred parts per million. The nature of the organic matter may be quite varied, depending on whether the organics are natural or synthetic in origin. Natural organics arise from the decay of vegetation from soil contacted by the water source. The products from the decaying process are rather complex but are usually acidic in character and polymeric in nature. Classification of the polymeric organic structures has been made in generic terms due to the complexity of the structures. The organic substances originating from decaying vegetational sources have been generically termed as humic and fulvic acids, based primarily on solubility considerations. Tannins, tannic acids, proteins, fatty acids, and polyols are among some of the less complex structures which have been specifically identified.

Synthetic organics may originate from a variety of sources, such as from the waste effluents of industrial operations and/or the discharge of domestic sewage. Many of the organic materials impart color to the waters and are eliminated only with great difficulty using conventional techniques.

Organic matter found in water supplies is usually acidic in nature and accordingly can be adsorbed by anion exchange resins. For most surface type waters, where organic matter is usually found, pretreatment of the waters can significantly reduce the organic content. Through the use of coagulation and clarification, organics present as turbidity (colloids) can be readily eliminated. Some of the soluble organics may also be reduced when organic polyelectrolytes (flocculants) are used in conjunction with an inorganic type coagulant. Chlorination of the waters does bring about some destruction of organics, but the prime purpose of chlorination is to render the biologically active organics innocuous.

Organic scavenger resins, such as Ionac® A-642, have, for example, been found useful for eliminating organic substances from a water supply prior to conventional demineralization. Organic scavenger resins commonly employed for this purpose are strong anion macroporous resins especially designed for this type of operation. Scavenger resins characteristically possess a highly porous structure and are capable of reversibly sorbing the organic species. That is to say, the organic matter, once adsorbed by the resin, can be quantitatively eluted from the resin under proper regeneration conditions.

It is customary to operate scavenger resins in the salt (chloride) cycle and to regenerate the resins with a warm brine solution. In many cases, for both economical and ecological reasons, the waste regenerants from the resin demineralization system can be effectively used to regenerate the scavenger resins, with or without a final short regeneration with fresh brine solutions.

TABLE 1

Organic Capacity — Ionac® A-642 Scavenger Resin

Organic Content (ppm as $KMnO_4$)	Service flowrate (bed volumes per hour)	Capacity (pounds of Organics, $KMnO_4/ft^3$)
1—10	16—20	0.6—0.8
10—25	16—20	0.8—1.0
25—50	12—16	1.0—1.2
50—100	6—10	1.2—1.5

Where a new demineralization system is being considered, a suitable macroporous anion resin can be employed instead of a conventional gel-type anion resin, eliminating altogether the need to use a scavenger unit prior to the demineralization train. This practice is recommended only in cases where the organic content of the clarified water supply to be treated is relatively low in soluble organics. Foregoing the use of a scavenger resin in favor of a macroporous strong base resin in the demineralization train should generally be considered only when the organic matter in the water supply is less than 30 ppm as $KMnO_4$ (i.e., this is the permanganate necessary to oxidize organic matter).

It is difficult to predict what organic loading capacities can be achieved using a scavenger resin because of the undefined nature of the organics present in the water supply. Nevertheless, approximate loadings that may be expected as a function of the organic content of the water supply are given in Table 1.

Regeneration of the scavenger resin is best achieved using a 10% NaCl solution at a 15 lb/ft^3 NaCl regeneration level applied at a two bed volume per hour flowrate. It is recommended that the salt regenerant be used at 120°F, particularly if the organic content exceeds 50 ppm as $KMnO_4$. For more aggressive waters and where inorganic colloidal compounds may also have been removed by the scavenger resin along with the organic matter, it is recommended that up to 2 lb NaOH per cubic foot of resin be incorporated into the salt regenerant solution.

The removal of organic tannins from residential water supplies cannot be accomplished by standard water filters and water softeners which employ a strong cation-exchange resin. When tannins are present in the water supply, they impart to the water a yellow or brownish tinge. A strong macroporous anion resin, such as Ionac A-642, can be added to the water softener unit to remove the tannins while the water is being softened. The same brine solution used to regenerate the softener resin will simultaneously regenerate the Ionac A-642 scavenger resin.

B. Phenol Removal

Wastewaters containing phenols present a serious disposal problem from an ecological standpoint. Though quaternary ammonium type anion-exchange resins have a good capacity for phenols, weak base anion-exchange resins of the polyamine or tertiary amine type are preferred since they can be regenerated with either dilute alkali or solvents such as low boiling alcohols.[1]

The sorption of phenol by weak base anion resins, based upon a styrene-divinylbenzene matrix, appears to proceed through two distinct mechanisms, depending upon the type and the ionic form of the ion exchange resin involved. For quaternary ammonium exchange resins in the salt form, the mechanism clearly does not involve ion exchange, since no anions are displaced from the resin during the adsorption of phenol.

TABLE 2

Phenol Capacity — Ionac® AFP-329

Influent phenol	Capacity (lb/ft^3, phenol)
100 ppm	0.22
500 ppm	0.62
1000 ppm	1.28
4000 ppm	3.60

This is supported by the fact that phenol may be eluted from the anion-exchange resin readily with an alcohol. This conclusion was also proposed by Anderson and Hansen[2] who, after determining the sorption isotherm for phenol on Dowex® 1-Cl, found the phenol isotherm curve to have the usual characteristics associated with sorption observed on common porous sorbents. Phenol sorption increased rapidly at low concentrations and approached a limiting value as the concentration of solute approached the water miscibility limit.

In the case of strong base anion-exchange resins in the hydroxide form, the phenols are removed primarily by an exchange for hydroxide ions. Under normal circumstances, regeneration of a quaternary amine exchanger to the hydroxide form is such that the resin functional sites will exist partially in the hydroxide form and partially in the salt form. The relative ratio of the two existing forms after regeneration depends, of course, on the caustic regeneration level. Under such conditions, two concurrently different sorption processes will be occurring, one involving adsorption of phenol at the salt functional sites, the other an interaction at the functional sites involving phenol and the hydroxide ion to form the phenolate ion.

Sorption of phenol on the weak, free base form polyamine anion-exchange resins appears to proceed mainly through an adsorption process, at least in the case of polyamine anion-exchange resins which do not contain quaternary ammonium functional groups. Waste aqueous streams and process streams containing phenol levels between 100 and 4000 ppm phenol can readily be processed yielding treated solutions containing less than 1 ppm phenol.

The working capacity of the resin for phenol is greatly dependent upon the initial concentration of phenol in the influent to be treated. In general, the capacity increases rapidly with an increase in the phenol content of the water to be treated. Typical capacities that can be realized using, for example, the weak base anion-exchange resin Ionac AFP-329 are listed in Table 2.

It should be noted that the phenol capacity of the Ionac AFP-329 resin is not adversely affected if the wastewater contains a neutral salt background, since weak base resins will not split salts to any appreciable extent. On the other hand, the presence of free mineral acidity in the influent may result in a reduced phenol loading. Likewise, the presence of alkali in the wastewater will also have the effect of reducing the phenol capacity.

The phenol sorption operation is carried out at flow rates between two to eight bed volumes per hour. A lower flow rate is beneficial for treating phenolic influents containing higher levels of phenol. In the case of a phenol influent containing 4000 ppm phenol (pH = 4.0), a flow rate of two to four bed volumes per hour results in a capacity of 3.6 pounds of phenol per cubic feet of resin to a 1% breakthrough while giving an average leakage of less than 1 ppm phenol.

Regeneration of the weak base ion-exchange resin is achieved using either a solvent such as methanol or isopropanol or a dilute NaOH solution. The use of alcoholic solvents allows for the recovery of phenol (as phenol) since the alcohol can be stripped from the regeneration waste stream, leaving behind a phenolic concentrate. The recovered alcohol can then be reclaimed and used for subsequent regenerations, while the phenol can be recycled back into a process stream. Use of dilute NaOH allows for the recovery of the phenol as the sodium phenolate salt.

C. Enzymes and Proteins

Ion-exchange resins can be used to recover and purify a variety of substances which may be difficult to obtain in a pure form using conventional techniques. One particular area of interest is the use of macroporous ion-exchange resins for the fractionation and purification of enzymes and related proteins[3] through the adsorption of the substances by the ion exchange resins.

The amphoteric character of the proteins allows them to behave as cations or anions, and they are therefore capable of being adsorbed by either cation- or anion-exchange resins. Elution of the peptides, once adsorbed, can be carried out by using acid, base, salt, or buffer solutions. In essence, an effective eluent will be one which reduces or eliminates the electrostatic forces binding the peptide or protein to the ion-exchange resin phase by either altering the charge carried by the peptide substance or by alternatively reducing the degree of ionization of the functional groups of the resins.

Although there are innumerable variations, present practice for the isolation of pure enzymes and other proteinaceous substances has been to carry out the concentration and preliminary fractionation of crude extracts by precipitation methods followed by gel filtration and chromatography in conjunction with ion exchange resins. Chromatographic techniques involving the use of ion-exchange resins have, within recent years, been introduced into many areas of biological research. One problem, in particular, that has received much attention is the purification of enzymes and their separation from other associated materials. Urease, trypsin, chymotrypsin, and pepsinogen have all been obtained in pure form by using ion exchange techniques.

Paleus and Neilands[4] were among the earliest investigators to use the carboxylic type cation exchange resins for protein chromatography applications. By virtue of its high isoelectric pH, cytochrome C was found to be firmly bound, below pH 9, to the ammonium salt form of carboxylic type weak cation exchange resins. It was also found possible to elute the cytochrome C by raising the pH to 10.8 with 0.1*M* ammonium hydroxide-ammonium acetate buffer. Carboxylic type weak cation-exchange resins have also been used extensively by other investigators, and many significant accomplishments have been reported in the literature involving the adsorption of proteins. Sulfonic type cation-exchange resins have also been used widely. An elution development technique employing Dowex 50 was used by Moore and Stein[5-7] for the fractionation of amino acid mixtures. Sodium citrate, sodium bicarbonate, and sodium carbonate buffers were employed over a pH range of 3.4 to 11.0. Although the capacity of the resin was found to be small, the resolving power was such that all the amino acid present in the protein hydrolyzates could be separated using very fine resin (200 to 400 mesh) and low flow rates. Chromatography employing anion-exchange resins has also been used primarily to fractionate acidic peptides.[8-11]

Since peptide separations by ion exchange usually require the use of buffers or salt solutions for elution, it is necessary to desalt the eluate fractions before recovering the peptides. In other instances, specific preparations require deionization before they can be processed. Mixed bed resins have been found useful for deionization purposes. For instance, Gillespie et al.[12] were able to use mixed bed resins for the deionization of *Aspergillus oryzae* prior to filter paper electrophoresis. Roboz et al.[13] used a mixed bed to effect the removal of salts from the pectin depolymerase obtained from *Neurospora crassa.* The earlier peptide separations of Moore and Stein and their collaborators used sodium acetate and sodium citrate buffers for elution. To avoid the necessity to desalt the individual eluate fractions before recovering the peptides, Margoliash and Smith[14] introduced volatile eluent buffers that could be removed by direct lyophilization. They were able to separate the peptides from a chymotryptic digest of cytochrome C by elution with pyridine-acetate buffers from a column of the pyridinium form of a sulfonic acid cation exchange resin of low cross-linking.

Basically, the mechanism by which a peptide (P) is adsorbed by a cationic or an anion exchange resin, may be visualized as follows:

$$R{-}COO^- Na^+ \quad + \quad NH_3^+P \rightleftharpoons RCOO^- NH_3P + Na^+ \tag{1}$$

$$R{-}NR_3^+ Cl^- \quad + \quad PCOO^- \rightleftharpoons R - NR_3{}^+PCOO^- + Cl^- \tag{2}$$

$R{-}COO^-Na^+$ and $R{-}NR_3^+Cl^-$ refer, respectively, to a carboxylic weak cation-exchange resin and a quaternary ammonium strong anion-exchange resin, where the adsorption of the peptide (P) solute by the ion-exchange resin substrates is influenced by several factors. The ion-exchange resins must have favorable pore characteristics when hydrated, so that no physical restrictions interfere with the diffusion of the relatively large organic molecules into the resin structures; favorable environmental chemical conditions must also exist. Since peptide substances are basically all amphoteric in nature because of their amino acid composition, the pH at which the adsorption is conducted will have a significant influence upon its degree of dissociation, its cationic or anionic behavior, and the solubility of the peptide itself under the given set of conditions. Consequently, the pH of the system is a dominant factor in determining whether a cation- or an anion-exchange resin can be best utilized as the adsorbent substrate for the organic substances.

With increased pH, the protein will be predominantly anionic in character; conversely, with a decrease in pH, the protein will be dissociated primarily as a cation.

$$\underset{\text{low pH}}{H_3{}^+N - R - \overset{\overset{\displaystyle O}{\|}}{C} - OH} \;\underset{(+)\,H^+}{\leftarrow}\; \underset{\text{isoelectric pH}}{H_3{}^+N - R - \overset{\overset{\displaystyle O}{\|}}{C} - O^-} \;\underset{(+)\,OH^-}{\rightarrow}\; \underset{\text{high pH}}{H_2N - R - \overset{\overset{\displaystyle O}{\|}}{C} - O^-} \quad + H_2O \tag{3}$$

Below the isoelectric pH, molecular association of the proteins through intermolecular bridging at the carboxylic acid portion of the protein is a possibility and can result in increased adsorption capacities. Molecular association of proteins at a pH below the isoelectric pH of the protein is probably at least partially the explanation of why carboxylic cation-exchange resins exhibit considerably greater adsorption capacities than anion-exchange resins when used to adsorb protein molecules from pH systems lower in pH than the isoelectric pH of the protein and at pH values where the carboxylic acid groups of the ion-exchange resin and the protein are undissociated.

In the case of weak electrolyte ion-exchange resins, the pH of the system will also influence the degree of ionization of the ion-exchange resin and consequently the ion exchanger's capacity for the organic substance under the adsorption conditions. The most unfavorable conditions for sorption of a solute by the weak electrolyte ion exchange resins will be when the adsorption is attempted at a pH which is in the vicinity of the dissociation constant (pK) of the ion exchanger and also close to the isoelectric pH of the solute. The ionic strength background of the system also influences the selectivity of resin for the organic solute. In general, with increased ionic strength of the solution phase, a decreased resin capacity for the protein solute will result due to the selectivity of the resin phase for other ions present in the solution.

Strong acid cation-exchange resins and strong basic anion-exchange resins should be effective in adsorbing a peptide solute over a broader pH range because their degree of ionization will not be as pH dependent as in the case of the weak electrolyte ion exchange resins. Consequently, the strong electrolyte resins will not be as useful as the weak ion exchange resin for the separation of solute mixtures characterized by isoelectric pH values a few pH units or less apart.

Elution of the solutes from the ion-exchange resins can be effected by using acidic, basic, or salt solutions; certain solvents such as alcohols may also be employed. In general, the eluent must be able to alter the degree of ionization of the active fixed functional groups of the ion exchange resin or must be capable of neutralizing the electrostatic forces binding the organic solute to the ion exchange resin substrate phase. Alternatively, the organic solute can be eluted from the substrate using a volatile acid or base such as acetic acid or ammonia and then recovered in relatively pure form after the volatile eluent component is removed from the eluate. Where nonvolatile eluents are used, these may be removed from the eluate by deashing with conventional gel-type mixed-bed resins, which will not adsorb any of the relatively large molecular weight organic solutes.

D. Antibiotics

Although it had been recognized for many years that ion-exchange resins could be used for the recovery and purification of a variety of biological products, it was not until the 1950s that the first ion-exchange process of this type went into commercial operation. This involved the recovery and purification of streptomycin using a carboxylic macroporous cation-exchange resin. Shortly thereafter, a similar operation followed for the recovery and purification of neomycin. In both operations, the antibiotic, present in unclarified fermentation broths, is adsorbed onto the carboxylic cation exchange resin, which is operated in the sodium cycle. After the resin becomes saturated with the antibiotic organic solute, the antibiotic is readily eluted from the resin using either dilute mineral acid, in the case of streptomycin, or with dilute ammonia solution, in the case of the neomycin. Following the elution step, the resin is converted back to the sodium form using a dilute sodium hydroxide solution, so that the resin can function again for the uptake of additional antibiotics.

E. Alkaloids

Though the adsorption of alkaloids by cation-exchange resins had been known for many years,[15] the commercialization of this technique for alkaloid recovery and purification was retarded until the development of synthetic acid-resistant cation exchange resins in the mid-1930s.[16,17] Riley[18] disclosed the use of these cation exchangers operated in the hydrogen cycle for the adsorption of nicotine. Higgins[19] also found cation-exchange resins operated in the sodium cycle to be effective for adsorbing nicotine. In addition, Higgins also found that strychnine, quinine, cinchonine, and other alkaloids could be adsorbed onto sodium form resins from alkaline, neutral, and even acid solutions. Tiger and Dean[20] were issued a patent for a process in which the hydrogen ions of a cation-exchange resin were displaced by alkaloid cations and the liberated nicotine subsequently recovered as an aqueous solution of its salt.

F. Fermentation Broths

The use of ion-exchange resins for the recovery of valuable organic substances from fermentation broths has been increasing in recent years. As an illustrative example, Jacob[21] discusses the recovery of xanthylic acid from fermentation broths. The acid is initially adsorbed onto a strongly basic anion-exchange resin and then eluted with an aqueous acid solution. Tsao et al.[22] extracted itaconic acid from molasses fermentation liquor by adsorbing the itaconic acid onto an anion-exchange resin. The best results were obtained by first passing the fermentaion liquor (diluted three fold with water) through a column containing a strongly basic anion-exchange resin. The itaconic acid was subsequently eluted quantitatively from the strongly basic anion-exchange resins using 3*N* NaOH.

G. Formic Acid Removal

During the commercial production of formaldehyde from the oxidation of methyl alcohol, the formation of some formic acid unavoidably occurs. Usually formic acid is present at a level of ca. 0.1% and can be readily removed through the use of a weak anion-exchange resin deacidification operation. Formic acid column capacities on the order of 3 to 4 lbs formic acid per cubic foot of resin can be achieved. Acrylic type weak base resins give somewhat higher capacities than stryene-divinylbenzene type weak base resins and are preferred when the original formic acid present in the formaldehyde exceeds a value of 0.1%. Elution of formic acid from the weak base ion exchange resins is readily accomplished using aqueous solutions of either NaOH or NH_4OH.

H. Non-Ionic Detergents

The removal of non-ionic detergents from waste waters, sewage, or domestic water supplies can be readily accomplished through the use of strong base or weak base anion exchange resins though most of the published literature refers only to the use of weak base resins for this purpose.[23,24] Non-ionic detergent levels in most surface and ground water supplies do not usually exceed 0.5 ppm, but on occasion, the levels in some rivers rise above 1 ppm. In general, the non-ionic detergent level in domestic sewage varies from about 0.3 to 3 ppm.

Strong anion resins are used in the chloride form as supplied for this purpose. Using flow rates of four to eight bed volumes per hour, non-ionic detergent levels can be lowered to values below 0.1 ppm (as alkyl benzene sulfonate) with capacities on the order of 0.3 equivalents of ABS per liter of resin being achieved. Regeneration of the resin is readily accomplished using two bed volumes of a one-to-one mixture of acetone and 4% HCl followed by rinsing with water. The weak anion-exchange resins are particularly useful for treating acidic streams containing the non-ionic detergents, in which case capacities as high as 0.5 eq ABS per liter of resin are found. In the case of the weak base resins, regeneration is accomplished using a one-to-one mixture of acetone and 4% NaOH.

I. Purification of Sugar Juices

Another interesting area of application involving the adsorption of organic substances is the purification of sugar bearing juices, liquors, and syrups. Ion-exchange resins are widely used to decolorize these solutions during some phase of their purification. Anion-exchange resins are particularly effective for this purpose, since most of the color impurities are organic substances exhibiting acidic characteristics. Because the active functional groups of anion exchange resins are basic in nature, they are capable of attracting the color bodies and bringing about their adsorption. The color impurities can subsequently be eluted from the resins using neutral brine solutions; acidic or alkaline regenerants can also be effectively utilized for this purpose. Although the complete chemical composition of the colored organic species found in sugar systems has not been fully established, due in part to the complex nature of the organic substances, polyphenols, tannins, and anthocyanin type substances of varied molecular weight have been identified.

The choice of a particular type ion-exchange resin is dependent upon the nature of the sugar being processed and whether or not some degree of deashing (demineralization) is also required. In the beet and corn sugar industries, some deashing of the sugar solutions is usually desirable during the refining operation. In these cases, it is customary to subject the sugar solutions to an ion exchange treatment using a cation-exchange resin and a weak anion-exchange resin demineralization system. The weak anion-exchange resin usually is macroporous and functions both to remove acidity generated by the cation-exchange resin and to adsorb organic color contaminants from the sugar solutions being processed. At cane sugar refineries, some form of carbon is used to decolorize the sugar liquors being treated. If animal rather than vegetable carbon is employed, partial deashing of the sugar liquors also occurs. In most cases, the carbon operations at cane refineries are supplemented by the use of strong base macroporous anion-exchange decolorizer resins. These resins are operated in the chloride cycle using brine regeneration and serve to further decolorize the partially decolorized char liquors. This usually results in a more economical operation and assures that highly decolorized sugar liquors are continuously available during the refining operation.

III. SORPTION FROM NONAQUEOUS MEDIA

With the introduction of new macroporous cation- and anion-exchange resins in the early 1960s, new impetus was given to the utilization of ion exchange for operations that needed to be performed in media other than aqueous. Prior to the introduction of the macroporous ion-exchange resins, the conventional gel ion-exchange resins proved to be inadequate for most nonaqueous applications because of the unfavorable kinetics and resin physical breakdown involved. The relatively slow absorption rates of nonpolar solvents by conventional ion-exchange resins cause ions or organic molecules to diffuse rather slowly through the resin internal gel structure where most of the active exchange sites responsible for ion exchange and adsorption of organic substances are actually located. As opposed to conventional gel type resins, macroporous type ion-exchange resins possess a more rigid internal pore structure having true measurable porosity whether the resin be in a hydrated or dry state. Accordingly, macroporous ion-exchange resin will always have a larger proportion of readily accessible functional sites per unit weight or volume of resin available for utilization in the nonaqueous solvent system. As a result, greater selectivities and rates of exchange and/or adsorption are realized in their use than would be found in the case of conventional gel ion-exchange resins.

The greatest drawback in the use of conventional gel ion-exchange resins for adsorbing organic species from nonaqueous media, however, rests in the excessive degree of swelling that they undergo during regeneration cycles in which polar solvents or solutions need to be employed. Since the macroporous resins do not undergo such drastic volume changes under identical conditions, they are found to be extremely resistant to attrition brought about by alternate contacts with solvent systems of different polarities. As a result, the structural stresses due to excessive swelling and contraction cause conventional gel-type resins to rupture drastically with cyclic use, making them inadequate for operations to be conducted in media other than water or in media possessing a dielectric constant approaching the same magnitude as found for water or the low molecular alcohols.

Macroporous quaternary ammonium anion-exchange resins[25] effectively remove inhibitors such as hydroquinone (HQ), monomethyl ether or hydroquinone (MEHQ) and tert-butyl-catechol (TBC) from ethylenically unsaturated monomers such as from styrene and divinylbenzene. Regeneration of the anion-exchange resins was readily accomplished using methanol, thus allowing the anion-exchange resins to be used over and over again for adsorbing additional inhibitor from inhibited monomers. It is interesting to note that conventional gel-type resins contracted drastically in volume when contacted with the ethylenically unsaturated monomers and failed to adsorb any appreciable amounts of the inhibitors. In contrast, the macroporous anion-exchange resins adsorbed up to 0.03 mmol MEHQ per milliliter of resin and up to 0.25 mmol HQ per milliliter of resin.

Sulfur compounds in petroleum stocks are noted to have an adverse effect on the quality of petroleum products. Thiols are particularly detrimental because of their disagreeable odor, strong corrosive tendencies, and deleterious effect on product stability.[26] Increased use of high sulfur crudes has magnified the need for greater control of thiols, in particular, aliphatic mercaptans.[27] It has been found that macroporous strong base anion-exchange resins quantitatively remove octylmercaptan from iso-octane solutions using a continuous column operation.[28] Methanol-conditioned hydroxide form resin was used since earlier preliminary efforts demonstrated the salt form resins to be ineffective. Similarly, using the free base form of a macroporous weak base resin possessing tertiary amine functional groups gave excessive mercaptan leakage practically from the onset of a column operation conducted on a laboratory scale. Iso-octane solutions containing as much as 500 ppm octylphenol were used in the studies.

Phenol can also be removed effectively from nonaqueous media having low nonpolar characteristics using macroporous quaternary ammonium resins. For example, utilizing a chloride form resin, a phenol adsorption capacity of 0.22 mmol/ml resin can be obtained to a 10 ppm phenol breakthrough when treating benzene solutions containing 300 ppm phenol. Using a flow rate of 1 gpm/ft^3 (eight bed volumes per hour), the 0.22 mmol/ml capacity was approximately three times greater than obtained using the corresponding conventional gel resin. Elution of phenol using methanol was quite sharp with better than 98% of the phenol recovered in two bed volumes of methanol.[28]

Macroporous sulfonic or carboxylic cation exchangers function well in adsorbing low molecular weight amine, such as butylamine from benzene or similar type media. Since a major portion of the total nitrogen found in gasoline and petroleum stocks occurs as basic nitrogen, the use of ion-exchange resins appears to be well suited for reducing this type of basicity. Hydro-treating operations at oil refineries are rather effective for reducing nonbasic nitrogen components but quite ineffective toward the basic nitrogen portion. Since reforming of gasolines over platinum type catalysts is extremely sensitive to feed nitrogen content, maximum reduction in the total nitrogen content of the feed becomes highly desirable.[28]

In the case of butylamine removal from benzene, a butylamine throughput capacity of up to 1.7 mmol/ml of resin can be achieved when using benzene solutions containing as much as 1000 ppm butylamine.[28] Needless to say, numerous other applications exist where an organic solute needs to be removed from nonaqueous systems. For these applications, macroporous ion-exchange resins now offer an alternative approach to existing methods which may be inadequate in many cases.

IV. CONCLUSIONS

The ability to utilize ion-exchange resins for the reversible sorption of a variety of organic substances has expanded the application realms of ion exchange. With the introduction of macroporous ion-exchange resins, which possess a definite porous network whether hydrated or anhydrous, the potential field of ion-exchange resin utilization has been vastly expanded. The use of ion-exchange resins beyond conventional softening type applications has proceeded certainly well beyond what might have been envisioned when synthetic cation ion-exchange resins, capable of exchanging ions, were first synthesized by Adams and Holmes in the mid-1930s. For the purification, separation, and recovery applications involving organic solutes, increased use of ion exchange resins has been particularly noteworthy during the past several decades and should continue at a progressively faster rate during the upcoming years. The growth potential of utilizing ion-exchange resins toward solving some of the many, often complicated ecological problems remains a definite possibility. Future progress towards this end is limited only by the imagination and ingenuity of the technician and investigator.

REFERENCES

1. Pollio, F. X. and Kunin R.,*Environ. Sci. Technol.,* 1(2), 163, 1967.
2. Anderson, R. E. and Hansen, R. D., *Ind. Eng. Chem.,* 47, 71, 1975.
3. Pollio, F. X. and Kunin, R., Chem. Eng. Prog. Symp. Ser., 67(108), 66, 1971.
4. Paleus, S. and Neilands, J. B.,*Acta Chem. Scand.,* 4, 1024, 1950.
5. Moore, S. and Stein, W. H., *J. Biol. Chem.,* 192, 663, 1951.
6. Moore, S. and Stein, W. H., *J. Biol. Chem.,* 211, 893, 1954.
7. Moore, S. and Stein, W. H., *Adv. Protein Chem.,* 11, 191, 1956.
8. Osterberg, R., *Biochim Biophys. Acta,* 42, 312, 1960.
9. Rudloff, V. and Braunitzer G., *Hoppe-Seyler's Z. Physiol. Chem.,* 323, 129, 1961.
10. Schroeder, W. A., Jones, R. T., Cormick J., and McCalla, K., *Anal. Chem.,* 34, 1570, 1962.
11. Strid, L., *Acta Chem. Scand.,* 13, 1787, 1959.
12. Gillespie, J., Jermyn M., and Woods E., *Nature,* 169, 487, 1952.
13. Roboz, E., Barrett, R. W., and Tatum, E. L., *J. Biol. Chem.,* 195, 479, 1952.
14. Margoliash, E. and Smith, E. L., *J. Biol. Chem.,* 237, 2151, 1962.
15. Ungerer, E., *Kolloid-Z,* 36, 228, 1927.
16. Griessbach, R., *Angew Chem.,* 52, 215, 1939.
17. Smit, P., *Chem. Weekbl.,* 39, 375, 1939.
18. Riley, F. R., U.S. Patent 2,226,389, December 24, 1940.
19. Higgins, J., British Patent 489,027 July 11, 1938.
20. Tiger, H. L., and Dean, J. G., U.S. Patent 2,293,954, August 25, 1942.
21. Jacob, T. A., U.S. Patent 3,366,627, January 30, 1968.
22. Tsao, J. C. Y., Huang, T., Wang, P., Lin, K., *J. Chinese Chem. Soc. (Peking),* Ser. II, 2(3), 1964.
23. Oehme, C. and Brost, H., British Patent 1,079,330, August 16, 1967.
24. Abrams, I. M., U.S. Patent 3,232,867, February 1, 1966.
25. McGarvey, F. X., Meunier, V. C., and Pollio, F. X., U.S. Patent 3,247,242, April 19, 1966.
26. Arnold, R. C. and Launer, P. J., *Anal. Chem.,* 24, 1741, 1952.
27. Seyfried, W. D., *Chem. Eng. News,* 27, 2482, 1949.
28. Pollio, F. X., unpublished data.

Chapter 2

BASE RECOVERY FROM SPENT SULFITE LIQUOR

A. Himsley

TABLE OF CONTENTS

I. INTRODUCTION

The pulp and paper industry has done a great deal of work in ion exchange for the recovery of soluble base from the spent sulfite liquor blown down from the digesters. Some of the processes which have been developed will be described as they offer techniques applicable to other industries.

The majority of pulp mills today are designed to produce kraft paper which employs a calcium bisulfite-sulfurous acid mixture for cooking the wood chips in the digesters. The calcium bisulfite is produced by introducing burner gas — a mixture of nitrogen and sulfur dioxide formed by burning sulfur in air — into the base of a tower filled with pieces of limestone. Water trickling down the tower meets the upward flow of gas, resulting in the production of the calcium bisulfite solution.

It was known for many years that a better quality and greater yield of pulp could be produced by the use of a soluble base such as sodium or ammonia. The soluble base gave a whiter pulp from which high quality papers are made. It also does not cause scaling-up due to insoluble calcium precipitates. However, the percentage of mills employing soluble-base cooking liquors is low compared to calcium-base users. The reason for this is the low cost of limestone compared to the higher costs of ammonia or sodium hydroxide.

II. ION-EXCHANGE PROCESSES

Many schemes have evolved for recovering the soluble base from the spent sulfite liquor. Among these are some ion-exchange techniques which are worthy of mention here. For a wider coverage of the field, the reader is referred to a comprehensive literature survey of sodium-base sulfite recovery processes by Collins and Shick.[1]

The spent sulfite liquor which issues from a digester has a pH usually between 3 and 5. Ammonia-base liquors contain about 6 g/ℓ NH_4, while the concentration of sodium-base liquors is around 12 g/ℓ Na. Divalent ions are around 0.2 to 0.5 g/ℓ, depending upon the nature of the soil in which the wood is grown and the local water supply.

The aim in recovering the sodium from this liquor is to return it to the digester as sodium bisulfite suitable for another cooking operation. It is logical, therefore, that one should aim at loading the resin with sodium and then stripping it off with sulfurous acid.

The low pH of the spent sulfite liquor rules out the use of the normal carboxylic-type weakly acidic resins, and therefore, a strongly acidic resin was initially considered. However, since the sulfurous acid prepared from burning sulfur contains approximately 5% of the acid at a pH of about 1.2, it is too weak to strip the resin free of all metallic cations and to restore the resin to the hydrogen form. On the other hand, such an acid strength readily regenerates a weakly acidic resin.

A. Weak Cation Exchange

Hussain[2] worked with a phosphorous-type resin (pK ∿3) and was able to obtain an adsorption efficiency between 90 to 99% with a regenerant solution between 3 to 7% SO_2; the stripping efficiency was around 94%.

However, the amount of sodium taken up by the resin when equilibrium was reached was quite low — about 0.07 eq/ℓ. Since the feed contains about 0.5 eq/ℓ, the volume of resin regenerated per day would have to be seven times the volume of liquor treated, which would be impractical.

B. Strong Cation Exchange

1. **Abiperm process** — There was further work done with a strongly acidic resin using liquid SO_2 to prepare stronger solutions of sulfurous acid and counter-current regeneration. This is known as the Abiperm process[3] a joint development of the Abitibi Paper Company and the Permutit Company. The flow sheet of the process is shown in Figure 1.

However, this process has not been developed due to the interference of divalent ions. Calcium and magnesium ions are more strongly held by the resin than sodium, and sulfurous acid is not strong enough to strip them off. As a result, the resin has to

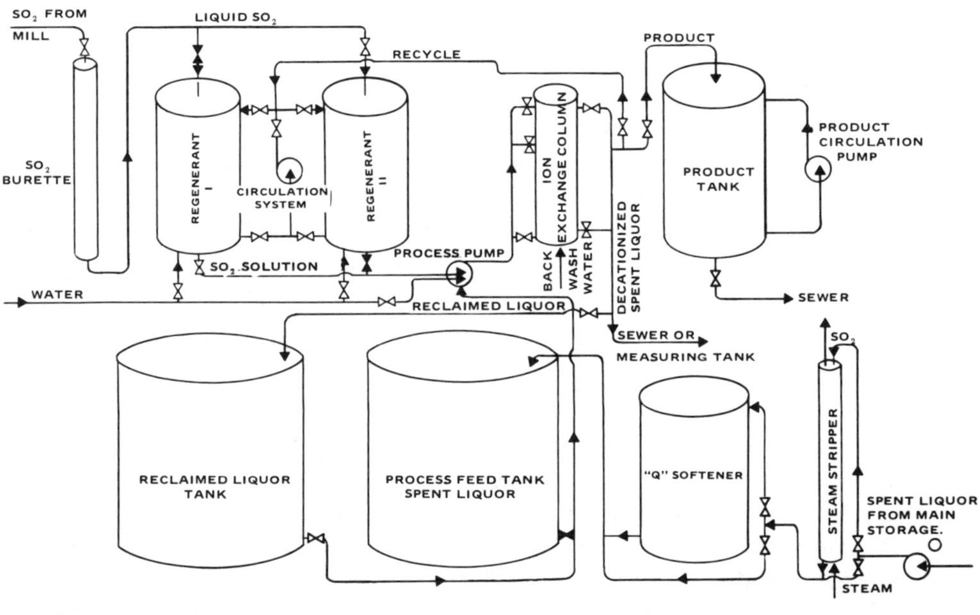

FIGURE 1. Schematic diagram of ion-exchange-base recovery plant (Abiperm Process). (From Manchester, D. F. and Termini, J. P., *Pulp Pap. Mag. Can.*, 62(9), T415, 1961. With permission.)

be periodically regenerated with a strong acid such as sulfuric or hydrochloric. It should be mentioned that because of the suspended matter in the spent sulfite liquor in the form of fibers, it is necessary to provide filtration ahead of the ion exchangers.

2. **Pritchard ORF process** — Another process, known as the Pritchard ORF process,[4] was developed by the Pritchard Company and the Ontario Research Foundation. This is shown in Figure 2. It is comprised of ion-exchange columns containing strong-acid cation resin. A strong solution of SO_2 is prepared under pressure at low temperature to strip the resin. However, in spite of expensive refrigeration and compression equipment, sodium recovery is poor and the removal of divalent ions from the resin is far from satisfactory.

3. **Pritchard Fraxon process** — Following closely upon this work, which was done at the Fraser Company Ltd. of Edmundston, New Brunswick, Wilson[5] found that a low pH solution could be prepared from burner gas with a ketone such as acetone as an additive. This solution readily eluted the divalent ions of calcium and magnesium from the resin and converted resin to the hydrogen form. This process is known as the Pritchard Fraxon process.[6,7] A flow sheet of this system is shown in Figure 3. The divalent ions that come off the resin along with the sodium are removed by precipitation and filtration. The acetone must, of course, be recovered to render the process economical, and a system using distillation for the recovery of acetone is shown in Figure 4.[6] It is also important to prevent the dilution of the eluted solution. The significance of this lies in the fact that because the spent sulfite liquor contains such a high concentration of soluble salts — as high as 12 g/ℓ Na — only about 3 bed volumes are treated before the resin becomes saturated. The volume of recovered acid employing a 50% recycle technique is 1.5 bed volumes, which shows less than a twofold concentration since recovery of soluble base is around 80 to 85%. Therefore, displacement and rinse water must be kept at a minimum in order to avoid excessive dilution with water. In 1969 it was estimated[7] that the installed cost of such a system for a mill using annually 18,500 tons sodium in cooking liquor would be approximately \$3,500,000.[6] Operating expenses were estimated at \$1.80 / ton of pulp. As yet, no mill has used this process.

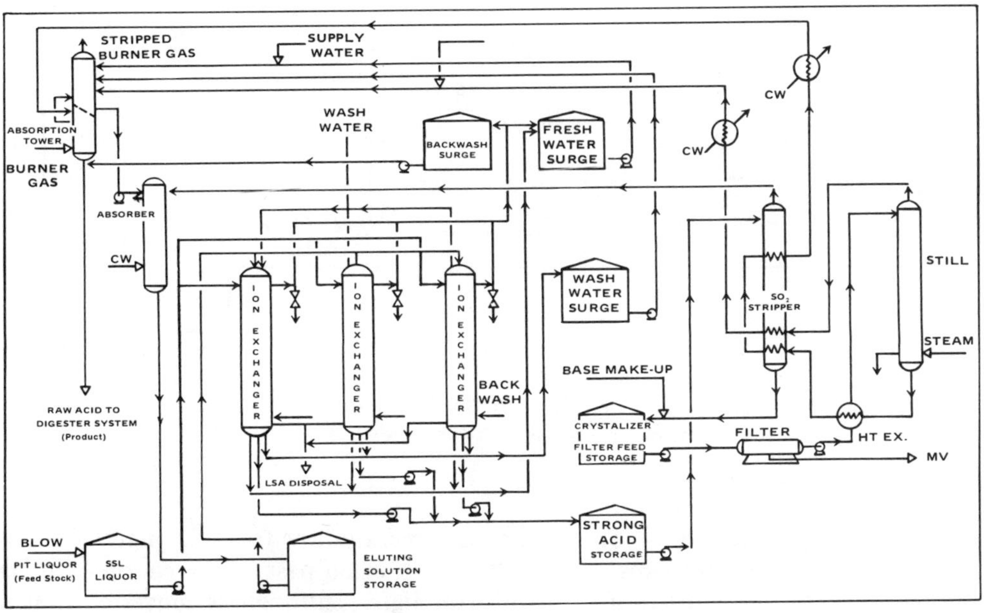

FIGURE 2. Flow diagram for sulfur-base recovery (Pritchard ORF Process).

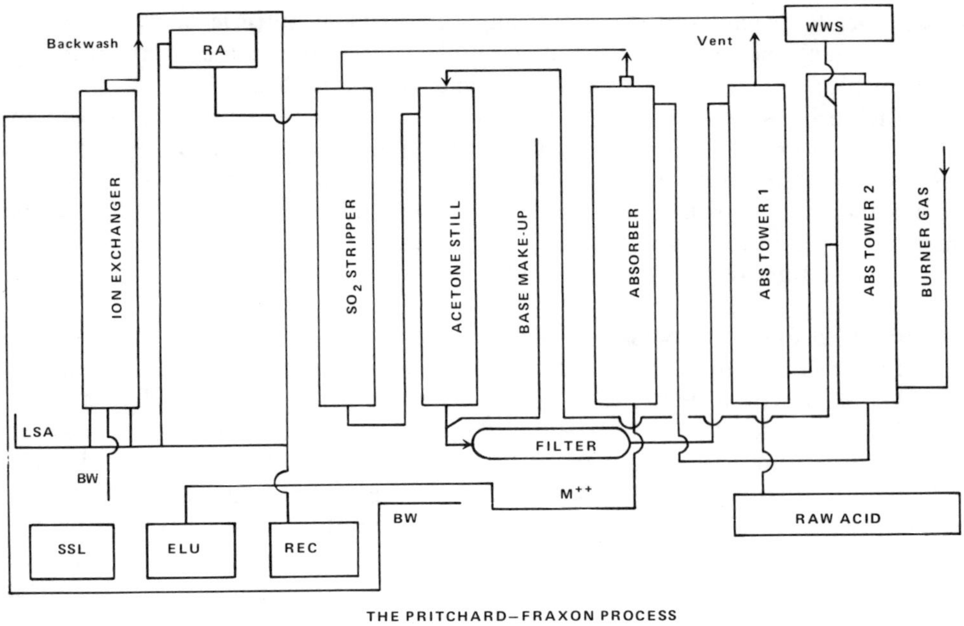

FIGURE 3. Flow diagram for Pritchard Fraxon Process. ABS, absorption; BW, backwash water; ELU, eluant; RA, recovered acid; REG, regenerant; REC, recycled eluant; SSL, spent sulfite liquor; WWS, waste water supply.

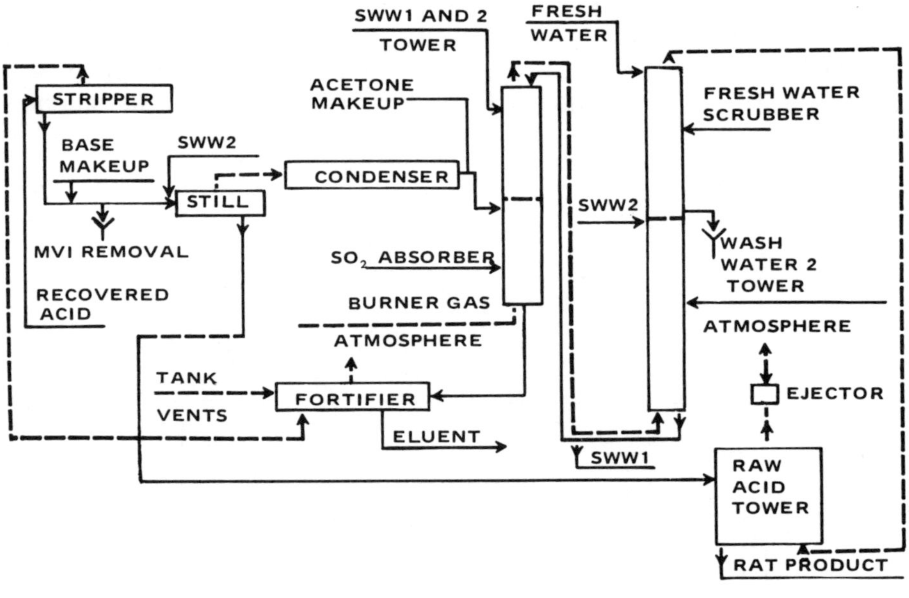

FIGURE 4. Flow diagram for acetone recovery. SWW = Spent wash water, MVI = multivalent ions, RAT = raw acid tower, and - - - = gas flow. (From Ali, S. F., Wilson, G. G., and Whitney, W. H., *TAPPI*, 51, 69a, 1968.)

4. **Lignin sulphonic acid** — The base-sulfite spent liquor from the digester is composed of base-lignin sulfonates (BLS). This solution on passing the cation exchanger in the H form becomes decationized, while the lignin sulfonate is converted to lignin

sulfonic acid as indicated by the following equations, where Equation 1 is decationization and Equation 2 is regeneration:

$$RH + B(LS) \rightarrow RB + H(LS) \quad (1)$$

$$RB + H^{+} X^{-} \rightarrow RH + BX \quad (2)$$

Lignin sulfonate has a high calorific value as a fuel when concentrated; also, it is used to a limited extent as a dispersant in boiler feedwater. Moreover, besides economic considerations, its elimination from the effluent reduces stream pollution which has plagued streams around paper mills.

III. SODIUM RECOVERY FROM ALKALINE WASTE STREAM

Another application of ion exchange involving pulp mill effluent is found at the Thorold Mill of the Ontario Paper Company.[8] At this mill, sodium is used as the soluble base, but the spent sulfite liquor is sent to an alcohol plant to reduce the fermentable sugars. From there, the wort goes to a vanillin process, and additional caustic is required to maximize production. The effluent is quite alkaline.

In this case, a process using a weakly acidic resin is used for taking up the sodium from the alkaline waste stream.[9] The sulfurous acid produced from burner gas is used to strip the metallic cations off the resin, resulting in a sodium bisulfite solution which is used in the digesters. The flow sheet is shown in Figure 5,[8] from which it is seen that three columns are employed, each of which contains approximately 340 ft^3 resin. Two units undergo elution concurrently, while one column is on the absorption step. Each column can complete an absorption and elution cycle in 2 hr. An automatic control system programs the units so that a continual stream of feed liquor and sodium bisulfite eluate is maintained to and from the system at all times.

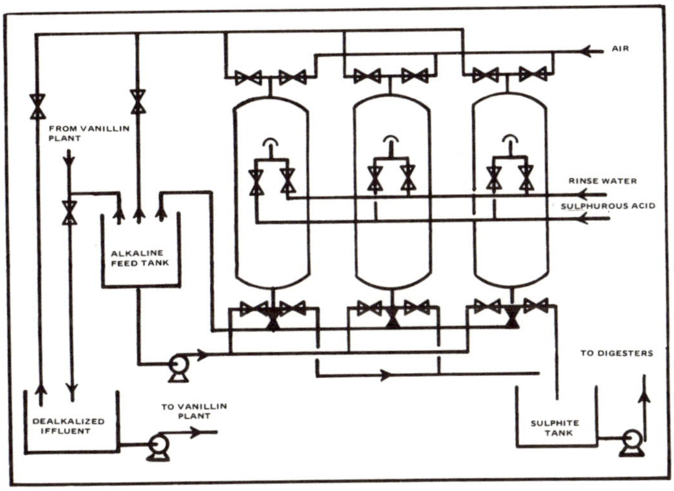

FIGURE 5. Flow diagram of ion-exchange system for sodium recovery.

COD, BOD, and color, they produce gelatinous sludges which are difficult to handle and dispose of and which have a high water retention so that filtration is difficult.

Poor decolorization is obtained when the effluent is treated by aeration. In lime precipitation, only those lignins with molecular weights higher than 350 produce precipitates.

Activated carbon, while effective, is too costly to be discarded after exhaustion, and regeneration results in the loss of carbonaceous material and a reduction in the capacity.

Ion exchange was investigated and found to be applicable and to offer many advantages which will be described later.[1-3] The ion-exchange process can complement other treatment processes of colored effluents in the wood industry as a posttreatment, e.g., after biological treatment where color reduction is limited.

The ion exchangers used are weak-base anion exchangers having good stability and resistance to fouling. The process requires three steps per cycle: (1) regeneration or activation, (2) service or purification, and (3) elution.

A. Regeneration (Activation)

A weak-base anion exchanger has an affinity for remaining in or going towards the base form. Therefore, the exchange must be carried out with the exchanger in the salt form $SO_4^=$ or $SO_3^=$. The equation for the regeneration reaction is

$$2R_3N + H_2SO_4 \rightarrow 2R_3\overset{\displaystyle H}{\overset{|}{N}}\text{—}HSO_4 \qquad (1)$$

B. Service Run (Purification)

The laboratory data showed that the exchanger has a higher affinity for the colored iogenic bodies than for the inorganic counter ions. This results in the release of the inorganic anions and the adsorption of the organic bodies during exchange. The equation for the service step can be written as

$$R_3\overset{\displaystyle H}{\overset{|}{N}}\text{—}HSO_4 + \text{Color body} \rightarrow R_3N \cdot \text{Color body} + SO_4^= \qquad (2)$$

Most of the SO_4 is left on the resin indicating that the adsorption is pH dependent.

TABLE 1

Color, COD, and BOD Produced in Various Stages of Pulp Bleaching

Pollutants/stages	A	B	Other	Total
Color (lignin)				
kg pt./ton pulp	1[a]	124	6	131
% of total	<1	95	4	100
COD				
kg/ton pulp	5.5	27	6.5	39
% of total	14	69	17	100
BOD				
kg/ ton pulp	4	6	2	12
% of total	34	50	16	100

[a] Normally, 5 to 10% of total.

C. Elution

Since the exchanger has an affinity for going towards the base form, the elution is done with a caustic or a basic solution. The equation for the reaction is

$$R_3N \cdot \text{Color body} + NaOH \rightarrow$$
$$R_3N + Na \cdot \text{Color body} + H_2O \qquad (3)$$

D. Other Operations

1. Rinsing of the bed is done after elution.
2. Backwashing is done often to remove the suspended solids or colloids.
3. The exchanged chloride during the service run can be removed with sodium sulfite or sulfate ions prior to elution.

E. Operational Parameters

The eluting agent is 4 to 8% caustic in strength and requires 15 kg per ton of pulp, depending on the influent color concentration. It was found that the basic solution known as oxidized white liquor may be used as the eluting solution.

The acid solution for the activation can be the effluent from A-stage, spent acid from the ClO_2 manufacturing plant, or an aqueous SO_2 solution. Thus, no new chemical solutions are required for the process.

To protect the ion exchanger, pretreatment must be carried out. This consists of sand filtration and trapping of the fibers.

The end point of the service run occurs when the color removal drops to 70% (i.e., when the color reaches about 400° Platinum (Pt) units; the average color removal is 90%). During the initial part of the service run or purification step, several bed volumes contain sodium sulfate. This can be saved in a holding tank for later removal of the Cl prior to elution. The colored bodies are not displaced because the affinity for the colored bodies by the resin is greater than for the inorganic ions.

III. COMMERCIAL APPLICATION

The Scandinavian plant consists of three or four units of beds. Two units operate in series during the run. One is the roughing unit and the second is the scavenging unit. When the roughing unit is exhausted, it is replaced with an activated or regenerated bed and the flow is reversed (i.e., the previous scavenging unit now becomes the roughing unit).

While the above units are in operation, the freshly removed unit is eluted with an alkaline solution. Upon completion of the elution step, one unit is regenerated with the acid solution from the A-stage. It was noted that the ion exchanger initially showed a drop of 15% in its capacity. The total plant capacity ranges from a flow of 100 to 240 m^3 of solution per hour. The process was developed by Uddeholms AB and one plant is located in Skoghall, Sweden. The process is licensed for the U.S. and Canada to Kamyr, Inc.

A. Performance

During a typical service run, the analysis of the effluent color, COD, and BOD on first alkaline extration had the following composition (see Figure 1):

Color	14000 units Pt (at pH 5.5)
COD	2400 ppm
BOD	800 ppm
Na	1500 ppm
pH	10.5

Table 2 shows the reduction in color, COD, and BOD in tons per day for a 300-ton pulp per day plant. It is to be noted that approximately 90% of the color is removed. Also, the removal of COD is high. While the degree of BOD removal is the lowest, in actual quantity it is the lowest of the several pollutants considered. Note: BOD/COD ratio increases, i.e., the result is a more biodegradable material.

In Figure 1 it is shown that 21 bed volumes of effluent from stage-B could be treated based on 30% color reduction. To obtain the most concentrated eluted solution, a very high concentration of eluting agent would be required. However, the concentration of the eluting solution is limited by the dosage. Analysis of the concentration of organic and inorganic material eluted is given in Figure 2, while the eluted color is given in Figure 3.

It is to be noted that most of the materials are eluted in less than 1.5 bed volumes, since the initial bed volume is interstitial water. It is possible to obtain a 20-fold concentration. This makes it possible to combine the eluate with weak black liquor from the brown stock washing system for evaporation and then incineration. Thus, there is no problem both in handling and disposing of troublesome sludge.

Another interesting point to be observed in Table 2 is the low level of removal of chloride; only 20% is absorbed by the exchange during the service run. The chloride in the influent to the exchanger can either be ionic or chlorine attached to organics. The fact that most of the chloride passes through the anion exchanger during the service run is of importance in that the eluate contains less chloride (80% is reduced), thus reducing the potential for corrosion in the evaporator.

The increase in organics available for incineration improves the energy balance. The average capacity is 270 kg (Pt) units of color per cubic meter of exchange.

B. Chloride Reduction

The B-stage effluent contains a considerable amount of chlorides carried over from the chlorination stage. The chlorides exist in two forms; most are ionic, and some become part of the organic entities.

If the exchanger is activated with $SO_3^=$ or $SO_4^=$, then most of the chlorides pass the resin bed with the decolorized effluent. Again, the selectivity variation of the exchanger for different ions plays an important role. It would appear that the selectivity is

TABLE 2

Degree of Color, COD, BOD, and Cl Removal in a 300 Ton Pulp Per Day Plant

	Influent			Effluent			% Removed	
Pollutant	B stage	Other stage	Total	B stage after ion exchange	Other stages not treated	Total	From B stage	From Total
Color (ton pt/day)	35	2.0	37	3	2.0	5.0	90	86
COD (ton/day)	6	3.5	9.5	1	3.5	4.5	83	53
BOD (ton/day)	2	2	4.0	1	2.0	3.0	50	25
Total Cl (ton Cl/day)	5	—	—	4	—	—	—	—
Dry matter (%)	0.6	—	—	—	—	—	15	—

$OH^- >$ colored organic ions $>$ (4)

Of course the concentration of an ion is an important factor affecting the affinity of the resin for a given ion. Chloride is not desirable in the evaporator, since it is highly corrosive.

To remove the chloride from the anion exchange prior to the elution of the organics, which later will be evaporated and incinerated, a sulfate or sulfite solution obtained in the first few bed volumes of the purification step is collected in a tank. Later, this solution is passed through the bed, resulting in the replacement of the inorganic chlorides by $SO_4^=$ or $SO_3^=$.

Figure 4 shows the organic chloride and ionic chloride in the eluate when the chloride is not displaced with the divalent ions.

C. Costs

The initial cost of the ion-exchange process for a 300 ton pulp per day was about one million dollars. When the plant produces more than 300 tons, another ion-exchange plant is necessary. However, today, with rising costs, a plant of 500 tons of pulp per day would cost three to six million dollars.

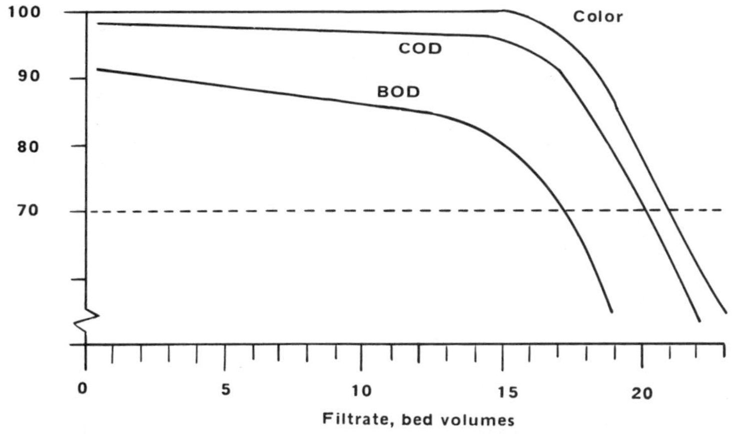

FIGURE 1. Percent removal in ion exchanger.

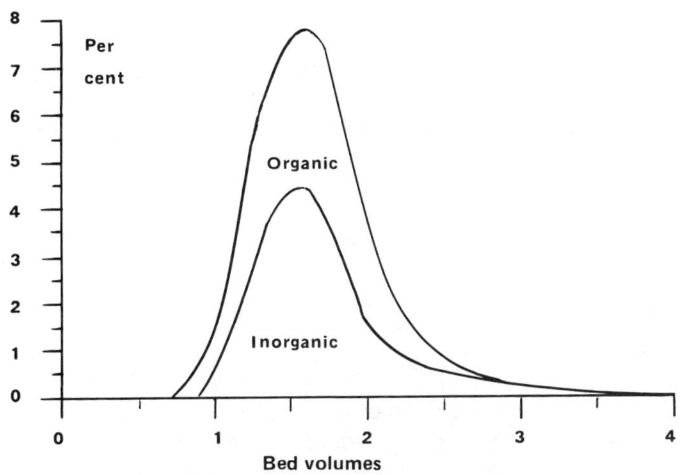

FIGURE 2. Dry substance in eluate.

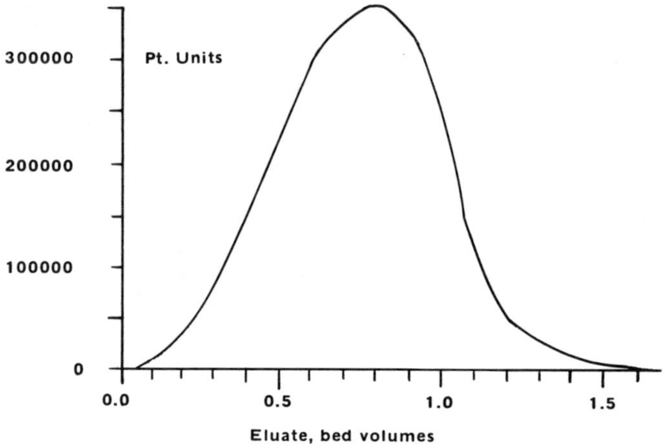

FIGURE 3. Color of eluate leaving the column.

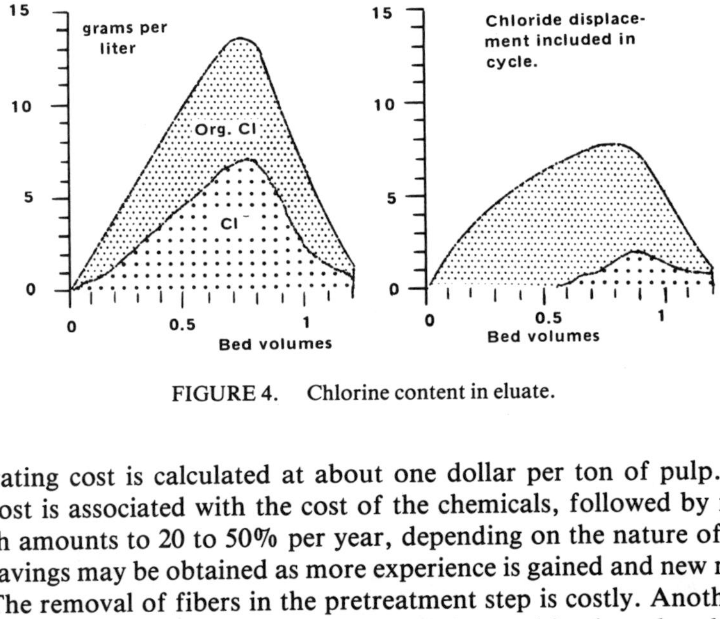

FIGURE 4. Chlorine content in eluate.

The operating cost is calculated at about one dollar per ton of pulp. Most of the operating cost is associated with the cost of the chemicals, followed by resin replacement, which amounts to 20 to 50% per year, depending on the nature of the influent. However, savings may be obtained as more experience is gained and new resins become available. The removal of fibers in the pretreatment step is costly. Another advantage is that complete counter-current washing is possible, resulting in a closed water system requiring only a limited amount of water for make-up.

In plants, where the effluent from the chlorination step is high in color, the effluent can be mixed with the first alkaline extraction effluent prior to the ion-exchange treatment.

IV. ADVANTAGES OF PROCESS

1. Almost complete decolorizing of effluents from the acid and first alkaline extraction stages.
2. Reduction in inorganic chloride ions going to the evaporator and incinerator.
3. Heat saving through concentration of the organics on elution.
4. Complete counter-current washing, which previously could not be done, since the first alkaline extraction effluent was too high in color and organics.

5. Dilute solutions can be treated very effectively.
6. It is a system free of sludges requiring dewatering, hauling, and disposing.
7. The chemicals required for the ion-exchange process are already available in the plant.

REFERENCES

1. **Anderson, L. S., Broddevall, B., and Lindberg, S.**, Color Removal from Effluents in Forest Industries by Ion Exchange, in Canadian Pulp and Paper Assoc. Meet., Montreal, Canada, 1973.
2. **Anderson, L. S., Broddevall, B., Lindberg, S., and Phillips, J.**, The Uddeholm Color Removal Process — a Field Report, in, Technical Assoc. Pulp and Paper Institute Meeting, Atlanta, Georgia, 1974.
3. **Anderson, K.**, The Uddeholm Ion-Exchange Process, Alkaline Pulping Conf., Dallas, September 13, 1976.

Chapter 4

REMOVAL OF ORGANIC MATERIALS FROM WASTEWATERS WITH POLYMERIC ADSORBENTS

E. C. Feeney

TABLE OF CONTENTS

I. INTRODUCTION

The primary contaminants found in chemical plant effluents are organic in nature and present in both ionized and non-ionized forms. It is well known that these contaminants can be selectively removed from water by ion exchange. However, significant levels of inorganic species are usually present in organic waste streams, and these species compete for the available ion-exchange sites. This causes a decrease in the resin's capacity for adsorbing organic materials. Another problem with this technique is that organic species are often exchanged or adsorbed irreversibly onto an ion-exchange resin, consequently causing a decrease in the operating life of the resin.

The use of polymeric adsorbents, a relatively new waste treatment technique, offers a more desirable alternative to ion exchange. Since polymeric adsorbents act as true adsorbents having no exchange sites, organic contaminants can be readily removed by regeneration. This chapter presents unpublished laboratory data.

II. POLYMERIC ADSORBENTS

A. Characteristics

Amberlite® polymeric adsorbents are hard, insoluble spheres characterized by a spectrum of surface polarities and by a variety of surface areas, porosities, and pore size distributions. An important aspect of the Amberlite adsorbents is the nature of their different surfaces. The phenomenon of adsorption on solids involves van der Waals' forces, which bind the adsorbate to a solid surface. Many types of interactions, such as hydrophobic bonding, dipole-dipole interaction, and hydrogen bonding, take place in the adsorption process.

While it is not possible to accurately predict just which materials will be adsorbed well by a given adsorbent, the general rule, however, holds that hydrophobic or non-polar molecules or portions of molecules are attracted to hydrophobic surfaces, and hydrophilic or polar material to hydrophilic surfaces. If each organic molecule is thought of as having both a hydrophobic and a hydrophilic end, then the hydrophobic end will be attracted to hydrophobic adsorbents such as Amberlite XAD-1, XAD-2, and XAD-4. This is particularly true when the adsorption takes place from an aqueous solution.

Adsorbents of intermediate polarity or hydrophilicity, such as Amberlite XAD-7 and Amberlite XAD-8, can have an attraction for either the hydrophobic or the hydrophilic end of a molecule. Cases will be found in which these adsorbents will function well in adsorbing both hydrophobic materials from water and hydrophilic materials from nonaqueous systems. For example, Amberlite XAD-7 is able to adsorb phenol, a moderately polar molecule, from either water or hexane with the adsorption taking place by association with the hydrophobic benzene ring during adsorption from water and the hydrophilic hydroxyl moiety during adsorption from hexane. Table 1 categorizes the Amberlite adsorbents in terms of polarity.

Another characteristic to be considered in the selection of the proper adsorbent to remove organic contaminants is pore size. Here the concept is very simple: The adsorbent must possess pores of a sufficient size so that the material to be adsorbed can migrate through the pore to the adsorbing surface. Table 2 gives the pore size of Amberlite polymeric adsorbents. It should also be emphasized that the number given is the average pore size and each adsorbent will possess pores of a range of sizes and, consequently, may adsorb considerably larger molecules than indicated by the average value.

A third important characteristic of an adsorbent is the surface area. For adsorbents of equal porosity, the greater the surface area, the greater the amount of solute that can theoretically be adsorbed. There is an inverse relationship between surface area and pore size with the smaller pore size adsorbents having greater surface area. As previously indicated, pore size is an important consideration in choosing an adsorbent for a particular application. This is illustrated by the observation that for a small molecule such as phenol, Amberlite XAD-4, which has a pore size of 90 Å, has a much higher capacity than Amberlite XAD-2, with a pore size of 50 Å, whereas for a larger molecule such as an alkylbenzene sulfonate, Amberlite XAD-2 has a much higher capacity than Amberlite XAD-4.

B. Regeneration

In order to use an adsorbent effectively, it is necessary to elute the adsorbed material. This can be accomplished in a number of ways (see Table 3). Some adsorbent users do not find it convenient to use organic solvents and therefore other methods of regeneration have been sought. Whenever ionic or potentially ionic materials are involved, it is generally true that materials will be much more strongly adsorbed in the

TABLE 1

Dipole Moments of Amberlite® Polymeric Adsorbents

Adsorbent	Moment
XAD-1, XAD-2, XAD-4	0.3
XAD-7, XAD-8	1.8

TABLE 2

Typical Properties of Amberlite® Polymeric Adsorbents

	Chemical nature	Porosity volume (%)	Wet density (g/cc)	Surface area (m^2/g)	Average pore diameter (Å)	Skeletal density (g/cc)	Nominal mesh sizes
Nonpolar							
XAD-1	Polystyrene	37	1.02	100	200	1.07	20 to 50
XAD-2	Polystyrene	42	1.02	330	90	1.07	20 to 50
XAD-4	Polystyrene	51	1.02	750	50	1.03	20 to 50
Intermediate polarity							
XAD-7	Acrylic ester	55	1.05	450	80	1.24	20 to 50
XAD-8	Acrylic ester	52	1.09	140	250	1.23	25 to 50

TABLE 3

Regeneration of Adsorbents

Regenerant	Application
Methanol or other organic solvent	Best regenerant for a wide variety of applications
Base	When adsorbate is weakly acidic
Acid	When adsorbate is weakly basic
Water	When adsorption is from an ionic solution
Hot water or steam	When adsorbate is volatile

non-ionic than in the ionic form. Thus, organic acid materials are best adsorbed from an acid solution where they are not ionized and are often eluted well with base, which ionizes the acid. Conversely, basic materials are best adsorbed from basic solutions and eluted with acid. Other techniques may find utility in special cases. If a material is adsorbed from a solution with a high salt background, it may sometimes be eluted with water. If one is dealing with a volatile material, steam may be an effective eluant. Table 3 summarizes the basic methods available.

The above concepts are best illustrated by a number of practical examples of the use of polymeric adsorbents in waste control. In some of these examples, it has been demonstrated that one can recover significant quantities of reusable chemicals that will more than pay for the waste treatment.

III. APPLICATIONS

A. Phenolic Wastes

Many industrial waste streams contain thousands of parts per million of phenol. This not only presents pollution problems but also constitutes economic losses of a valuable raw material. An ideal waste treatment process for such streams would remove phenol from the stream and recover it in a usable form.

1. Process Description

The aqueous waste stream containing phenol or similar compounds is passed through one or more columns of the polymeric adsorbent, yielding an effluent with a phenol content generally less than 1 ppm. The capacity of the polymeric adsorbent will depend upon the type and concentration of phenolics in the raw waste water and the background concentration of other organics and salts. A high salt background enhances phenol adsorption, probably due to a salting-out effect. Halogenated phenolics are adsorbed even more readily than nonsubstituted phenols. Since phenolic compounds are weak acids, their adsorption on the polymeric adsorbent is affected by the pH of the solution. The adsorptive capacity of the resin will change sharply with increasing pH, since the phenolic molecule will change from a neutral, poorly dissociated form at low and neutral pH to a highly dissociated anion form at high pH. The neutral species is highly adsorbed by the polymeric resin, while the more polar, charged form exhibits very little adsorption.

Specific adsorption data for low to moderate concentrations of phenol and substituted chlorophenols on a polymeric adsorbent are given in Table 4.

TABLE 4

Adsorption of Phenolic Compounds by AMBERLITE XAD-4

Compound	Influent concentration (ppm)	Resin capacity (lb/ft^3 at 0.1 ppm leakage)
Phenol	6700	5.4
Phenol	3000	4.5
Phenol (in 13% Nacl)	250	1.0
Phenol	250	0.8
m-Chlorophenol	350	2.4
2,4-dichlorophenol	430	5.1
2,4,6-trichlorophenol	510	12.0

Note: Flow rate = 0.5 gpm/ft^3; temperature, 25°C.

Column experiments are essential in any phenol application to select the best resin, confirm resin performance, and determine the optimum waste flow rate through the bed.

2. Regeneration

To regenerate the polymeric adsorbent, the attractive forces between the phenol molecule and the polymer must be overcome by either the solvating power of a polar organic solvent on the phenol molecule or transformation of the phenol molecule to a more soluble phenolate ion. Although the binding forces between phenol and the polymeric adsorbent are strong enough to allow the adsorbent to effectively remove phenol from the waste stream, the attraction is weak enough to make either solvent or caustic regeneration feasible. The most efficient solvent regenerants are low molecular weight alcohols and ketones such as methanol and acetone.

The choice of regeneration system will depend upon the needs and circumstances of each individual situation. The method selected, in turn, depends on the desired purity of the recovered phenol, availability of solvents (such as acetone or methanol) at the plant site, or the availability of solvent-phenol distillation equipment. In every case, the concentration of phenol in the regenerant stream will increase as the concentration of phenol in the raw waste stream increases. The following are five different regeneration schemes. It is our belief, based on field experience, that every waste phenol application can use one of these systems.

a. Dilute Caustic Regeneration

This system will find the greatest application in situations where direct recycle of a dilute sodium phenolate stream is desirable, or where waste caustic is available. In order to recover the phenol from the sodium salt, a sulfuric acid treatment is recommended if direct recycle is not feasible. If on-site recovery is not attractive, the sodium phenolate can be sold to firms specializing in recovery of this material.

b. Concentrated Formaldehyde Regeneration

This system will find application where a phenol-formaldehyde mixture can be directly recycled, such as to a polymerization reactor. By using 44% formaldehyde, the phenol can be efficiently eluted from the resin, and the entrained water in the bed will dilute the resultant mixture to 37% formaldehyde. A pilot plant has been operated for 140 cycles without evidence of resin fouling via polymerization.

c. Solvent Regeneration with Recycle of the Phenol-solvent Mixture

When recycle of the phenol in a water-wet solvent is acceptable or in applications where waste solvents are available, this system will find the greatest use. The solvents found to be most effective in phenol regeneration from Amberlite polymeric adsorbents are methanol and acetone. Two bed volumes of either solvent are usually effective in removing virtually all of the adsorbed phenol. Since phenol loadings vary with the concentration of phenol in the influent waste stream, the two bed volume regeneration will result in a phenol concentration of from 1 to 10% in the solvent. Note that water entrained in the bed is unavoidably present in the phenol solvent mixture but generally can be held to less than 25%.

d. Solvent Regeneration and Recovery with Recycle of the Aqueous Phenol Stream

Again, two bed volumes of solvent are used to regenerate the resin. The spent methanol or acetone regenerant is fed to a distillation unit where the solvent is recovered for subsequent regenerations. A phenol-rich water stream is taken off as a bottoms product. If desired, a second distillation step to recover solvent washed from the resin bed after the regeneration will increase solvent recovery.

e. Solvent Regeneration and Solvent Recovery with Recycle of the Concentrated Phenol Stream

The previous system is further modified by adding a separator and a distillation column to dehydrate the phenol-rich water stream to yield 98% phenol; a small 10% phenol/90% water azeotrope stream is recycled to the influent of the adsorption train. A flow sheet of this system for a continuous distillation operation is given in Figure 1. The solvent and phenol recovery operation could also be conducted in batches.

B. Trinitrotoluene (TNT)

Pilot-scale studies conducted at the Iowa Army Ammunition Plant, Burlington, Iowa, have demonstrated a safe, practical and economical polymeric adsorption process for removing trinitrotoluene (TNT) from waste streams. Once loaded with waste material, the adsorbent can be safely regenerated with solvent. To minimize operating costs, the solvent can then be easily recovered for reuse, leaving only a concentrated aqueous sludge of explosive contaminants for ultimate disposal. The use of carbonaceous material for removal of TNT is clearly not indicated, since regeneration can be achieved only thermally and such practices are clearly unsafe for a carbon saturated with an explosive such as TNT.

As a result of the success of this pilot-scale study, full-scale demonstration installations are being planned for several ordinance plants in this country. This same process is equally applicable to the removal of dinitro-toluene (DNT), nitrocresols, and other

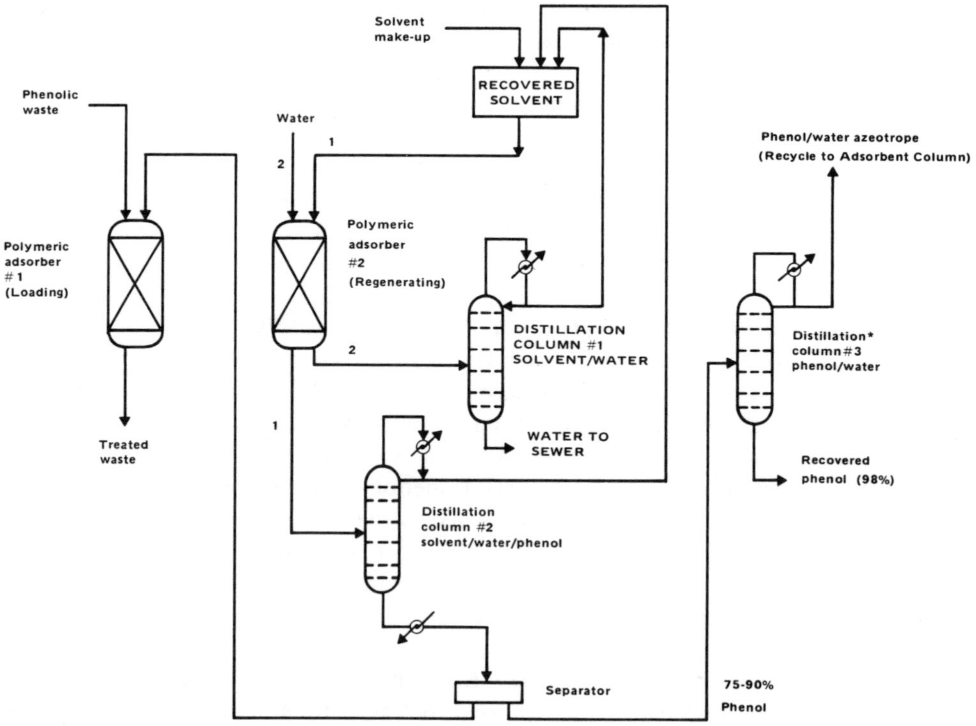

FIGURE 1. Phenol removal and recovery system for solvent regeneration of Amberlite® adsorbent. Column 3 is utilized when highly pure phenol is required.

hazardous explosive materials as well as to the treatment of industrial waste streams containing aromatic nitrobodies.

C. Chlorinated Pesticides

Amberlite polymeric adsorbents have also demonstrated applicability in the treatment of industrial waste streams containing chlorinated pesticides. Large amounts of laboratory data have been generated using model compounds of the type encountered in pesticide-manufacturing waste streams. Figure 2 presents leakage curves comparing Amberlite XAD-4 and granular, activated carbon run under identical conditions using a waste water obtained from a chlorinated pesticide manufacturer as influent. It is apparent that the polymeric adsorbent is superior to the activated carbon both in terms of pesticide leakage and in overall operating capacity. The regeneration of the polymeric adsorbent with isopropanol (shown in Figure 3) is efficient. Thermal regeneration, while more efficient for the carbon results in as much as a 10% loss of the activated carbon, significantly increasing the operating expense.

Based on the above studies, a treatment process utilizing the nonpolar Amberlite polymeric adsorbents has been installed to treat the effluent at a large pesticide manufacturer's facility. The system is operating successfully.

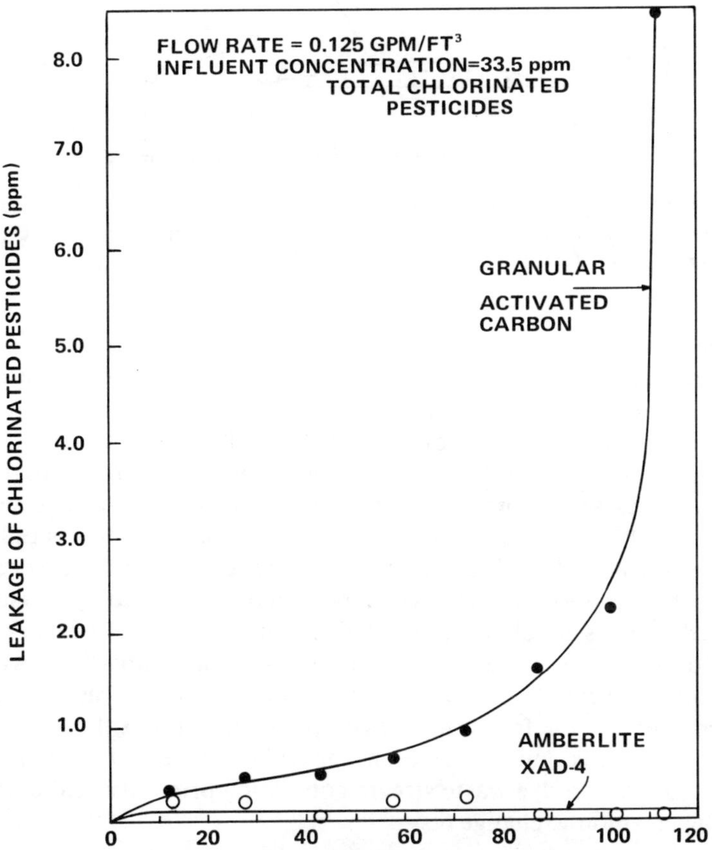

FIGURE 2. Adsorption of chlorinated pesticides with Amberlite® XAD-4 and a granular activated carbon. (Reprinted with permission from Kennedy, D. C., *Environ. Sci. Technol.*, 7, 138, 1973.)

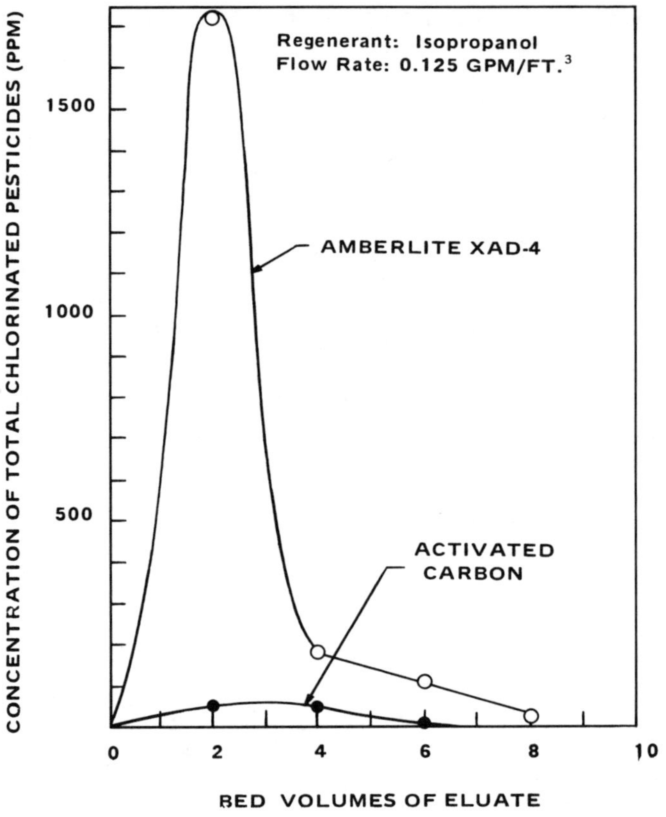

FIGURE 3. Regeneration of Amberlite® XAD-4 and a granular activated carbon with isopropanol. (Reprinted with permission from Kennedy, D. C., *Environ. Sci. Technol.*, 7, 138, 1973.)

D. Dye Waste

With the current trend in industry toward the use of more resistant dyes, conventional biological and chemical treatment techniques may have to be supplemented or replaced to meet more stringent color standards. While activated carbon is the material most commonly used to adsorb organic material, it has several limitations: thermal regeneration is necessary before placing the carbon back in service, generally resulting in losses of 5 to 10% of the carbon per cycle; regeneration removes only the organic components of the adsorbed dye waste; metallic residue coats the carbon surface and results in loss of adsorption capacity when treating premetallized dye wastes.

A combined process involving polymeric adsorbents and ion-exchange resins has been shown to be an economical waste treatment system. Amberlite XAD-7 polymeric adsorbent removes the bulk of the soluble organic from the waste stream. The ion-exchange resins remove the few ionic dye components that pass through the adsorbent. This process can be customized for each particular effluent. For example, the combined process can treat a dye waste stream containing primarily acidic or anionic dyes by substituting an anion-exchange resin.

A major dye producer has operated an adsorption-ion-exchange system to decolorize 300,000 gal of waste per day for over 2 years. The system reduces the average color of the waste stream from 75,000 to 500 American Public Health Association (APHA) units on the Pt-Co scale and chemical oxygen demand (COD) from 5280 to 2600 ppm.

It also removes copper and chromium present in the waste stream as both salts and organic chelates.

IV. SUMMARY

Polymeric adsorbents have already gained recognition as a viable separation technique for organic materials. These products are commercially available and are now being used in industrial processes. However, this development is still at the embryonic stage, with many investigators now evaluating the properties of the polymeric adsorbents in the laboratory and in pilot plants. While polymeric adsorbents are in competition with other commercially available adsorbents such as granular carbons, silica gel, and bone char for fluid treatment applications, their utility rests in their high capacity, selectivity, and regeneration efficiency. Along with these advantages are their outstanding chemical and physical stability that should yield a long service life.

Other Applications

Chapter 5

WATER TREATMENT IN NUCLEAR POWER PLANTS

G. P. Simon

TABLE OF CONTENTS

I. INTRODUCTION

Nuclear reactors create large quantities of potentially harmful radioactive nuclides. It is the general concern of reactor technologists to achieve some useful objective from the controlled fission process while at the same time rendering harmless the radioactivity produced. The primary coolant and moderator water in contact with the nuclear core acts as a vehicle of transport for the radionuclides produced in the core and reactor vessel surfaces to areas outside the reactor vessel and ultimately to the environment.

Radionuclides are produced by the alteration of the atomic nucleus followed by the emission from the altered or excited nucleus of energetic elementary particles or photons. The radioactivity, or decay, is usually designated by the elementary particles emitted. Table 1 is a list of these emissions and elementary particles of interest in reactor coolant technology including those involved in basic nuclear reactions.

These emissions are all basically harmful to biological living systems. The hazards may arise from the radioactivity acting either externally or internally. The relative hazard is dependent upon the emission, its chemical nature, and the decay rate of the radionuclide under consideration.

II. RADIOACTIVE DECAY

Radioactive decay is characteristic of a wide range of both naturally occurring and synthesized radionuclides. Decay rate is a unique property of each nuclide. This characteristic change can be described mathematically as follows:

TABLE 1

Radioactive Emissions and Nuclear Particles[1]

Emission	Charge	Mass no.	Symbol	Energy range (MeV)
Positron	1+	~0	$\beta+$	up to 4—5
Beta	1−	~0	$\beta+$	0.015—15
Neutron	0	1	n	up to 12
Alpha	2+	4	α	1.5—11.7
Gamma	0	0	γ	0.01—7
Proton	1+	1	p−	—
Deuteron	1+	2	D	—
Triton	1+	3	T	—

$$\eta = \eta_0 \, e^{-\lambda t} \qquad (1)$$

where η_o = number of radionuclide atoms present at time zero, η = number of radionuclide atoms present at any time, λ = characteristic decay constant, and t = time.

For all but very special circumstances and then only to a minor degree, λ is independent of temperature, pressure, and chemical state. The time ($t_{1/2}$) required for an initial number (η_o) of radioactive atoms to decay to half its value is defined as the "half-life" and is expressed by:

$$t_{1/2} = \frac{0.693}{\lambda} \qquad (2)$$

The characteristic half-lives of the known nuclides extend over a wide range, from fractions of a millisecond to up to 10^{15} years. The decay rate, or half-life, is an important factor in considering any waste treatment process, since the products being treated may involve both high-energy emission with a short half-life and low-energy emission with a relatively high persistence or long half-life.

Radioactivity is generally expressed as the rate of disintegration without regard to the type of emissions involved. The standard unit is a curie, and is defined as the quantity of radionuclide in which the number of disintegrations per second is 3.7×10^{10}; the weight, in grams M_{Ci}, of one curie of a single radionuclide is given by

$$M_{Ci} = 6.15 \times 10^{-14} \frac{M}{\lambda} \qquad (3)$$

where M is the gram molecular weight and λ is the decay constant expressed in reciprocal seconds. Table 2 gives a comparison of the curie weight of some commonly encountered radionuclides in primary coolant water from pressurized and boiling water reactors; an indication of the source of the radionuclide of interest is also included in this table.

The energy of radionuclide decay is, by the law of conservation of momentum, distributed between the emitted particle and the daughter nuclide. This distribution is dependent on the mass relationships between the radionuclide and the emitted particle. In beta decay, a high proportion of the energy is emitted with the particle. In alpha decay, a somewhat high proportion of the energy is retained by the daughter radionuclide due to the relatively large mass of the alpha particle. The energy distribution differences between beta and alpha decay are illustrated in Figure 1. This figure also shows the great variety available for energy release as the radionuclide approaches a stable ground state. In assessing the relative significance and treatment process to con-

TABLE 2

Half-life and Mass Relationship of Radionuclides[2]

Radionuclide	Source	Half-life	Mass (g/Ci)
$^{59}_{28}Ni$	Structural	8×10^4 years	13.1
$^{14}_{6}C$	Make-up water	5600 years	0.22
$^{137}_{55}Cs$	Fuel leak	33 years	1.26×10^{-2}
$^{90}_{38}Sr$	Fuel leak	28 years	7.03×10^{-3}
$^{60}_{27}Co$	Structural	5.27 years	8.8×10^{-4}
$^{55}_{26}Fe$	Structural	2.94 years	4.5×10^{-4}
$^{59}_{26}Fe$	Structural	45 days	2.03×10^{-5}
$^{134}_{55}Cs$	Fuel leak	2.3 years	8.6×10^{-4}
$^{106}_{44}Ru$	Fuel leak	1.0 years	2.95×10^{-4}
$^{144}_{58}Ce$	Fuel leak	282 years	3.1×10^{-4}
$^{58}_{27}Co$	Structural	71 days	3.14×10^{-5}
$^{51}_{24}Cr$	Structural	27.8 days	1.08×10^{-5}
$^{89}_{37}Rb$	Fuel leak	19.5 days	1.28×10^{-5}
$^{140}_{56}Ba$	Fuel leak	12.8 days	1.37×10^{-5}
$^{131}_{53}I$	Fuel leak	8.14 days	8.15×10^{-6}
$^{90}_{39}Y$	Fuel leak	61 hr	1.75×10^{-6}
$^{140}_{57}La$	Fuel leak	40 hr	1.78×10^{-6}
$^{140}_{57}La$	Fuel leak	40 hr	1.78×10^{-6}
$^{24}_{11}Na$	Make-up water	15.1 hr	1.15×10^{-7}
$^{35}_{17}Cl$	Make-up water	4.4×10^5 years	44.1
$^{35}_{16}S$	Make-up water	87.1 days	2.34×10^{-5}
$^{64}_{29}Cu$	Structural	12.8 hr	2.61×10^{-7}
$^{56}_{25}Mn$	Structural	2.6 hr	4.59×10^{-8}
$^{16}_{7}N$	Make-up water	7.4 sec	1.04×10^{-11}
$^{59}_{38}Sr$	Fuel leak	53 days	3.61×10^{-5}
$^{18}_{9}F$	Water	110 min	1.05×10^{-8}
$^{13}_{7}N$	Water	10 min	6.92×10^{-9}
$^{54}_{25}Mn$	Structural	313 days	1.29×10^{-4}
$^{133}_{53}I$	Fuel leak	21 hr	8.93×10^{-7}
$^{135}_{53}I$	Fuel leak	6.7 hr	2.88×10^{-7}
$^{95}_{40}Zr$	Structural	65 days	4.72×10^{-5}
$^{181}_{74}W$	Structural	130 days	1.8×10^{-4}
$^{187}_{74}W$	Structural	24 days	3.44×10^{-4}

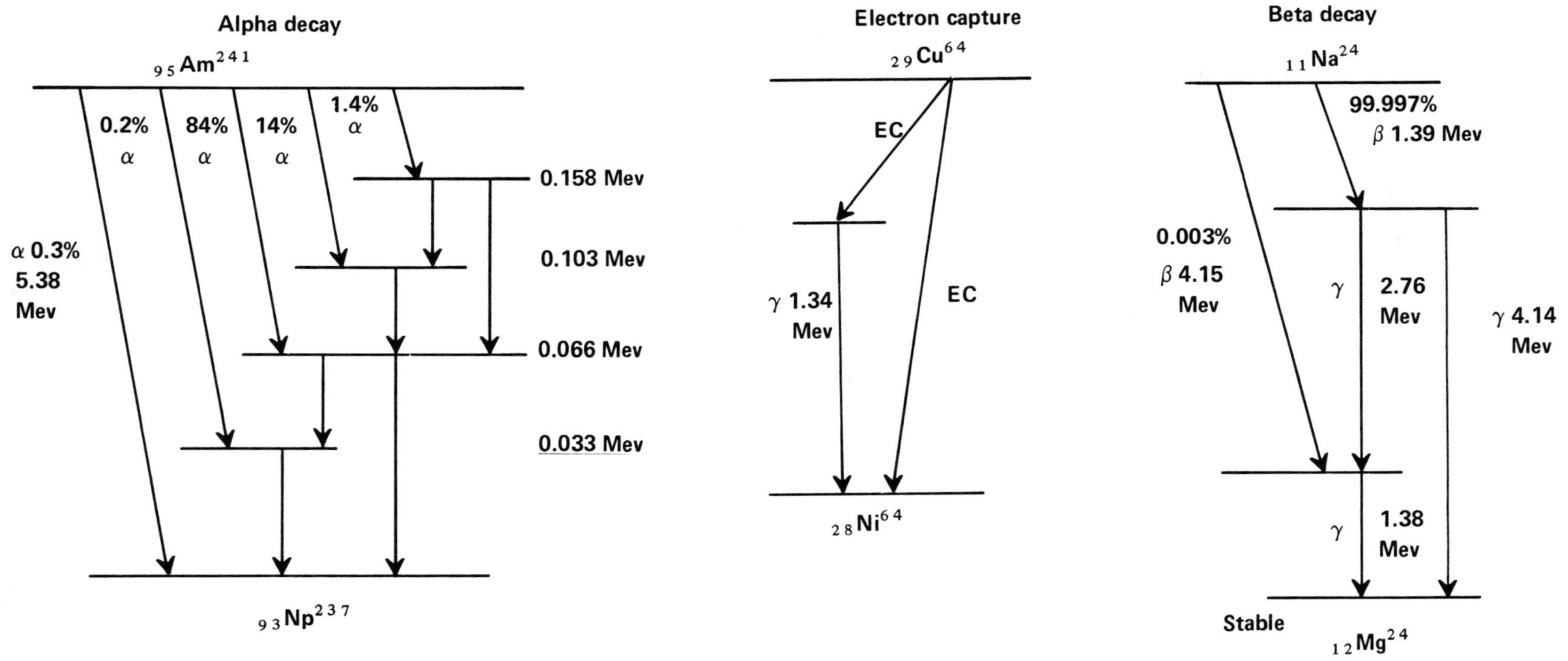

FIGURE 1. Radioactive decay schemes. (From Radiological Health Handbook, Environmental Protection Agency, Sanitary Engineering Center, U.S. Department of Health, Education and Welfare, 1954, 83. With permission.)

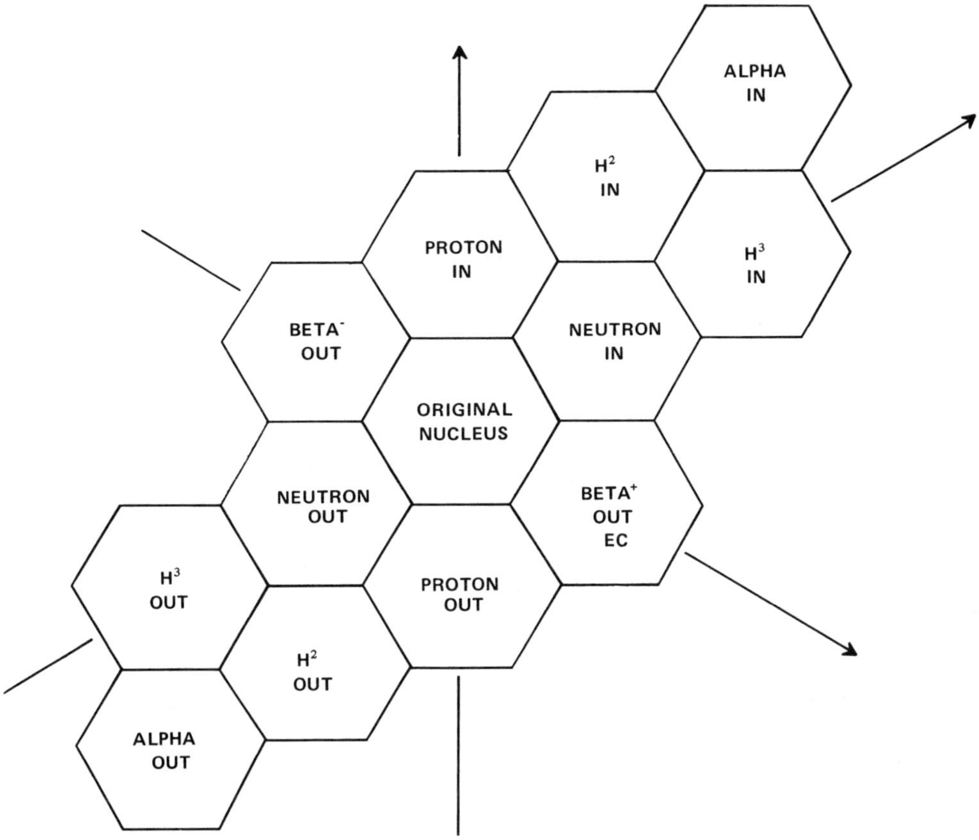

FIGURE 2. Trilinear chart of nuclides. (From Sullivan, W. H., *Trilinear Chart of Nuclides,* Oak Ridge National Laboratory, U.S. Government Printing Office, Washington, D.C., 1954, 83. With permission.)

trol a radionuclide, it is essential to review the decay scheme and evaluate its impact on the environment and the treatment process being used.

The chemical changes associated with radionuclide decay are readily seen on a chart of nuclides arranged according to constant number of neutrons (Z) and constant mass number (A). An excellent reference chart of this type is the Trilinear Chart of Nuclides, available from the Government Printing Office.[3] Figure 2 is a key to types of radioactive transformation available from this chart.

In the production of electric power by nuclear means, the predominant reactor system currently in use is the pressurized water reactor (PWR). In these systems, the heat is extracted from the core with water under high pressure, so as to keep the system below the nucleate boiling point. The water still under pressure is circulated in a closed loop through a steam generator back to the reactor core. In most systems of this type, a small percentage of the total flow is continuously demineralized and returned to the main stream. This by-pass purification is used to maintain the quality of the coolant water; in addition, it helps control the radioactivity levels formed in the water from corrosion and abrasion of internal surfaces wetted by the rapidly recirculating primary coolant. Radionuclides formed from corrosion products and other sources are shown in Table 3 for a typical PWR power plant.

The radionuclide distributions given in Table 3 will vary with the nature of the reactor structural materials, fuel cladding, chemicals added for corrosion control, water

TABLE 3

Typical Radionuclides in Coolant of a Pressurized Water Nuclear Power Plant[4]

Nuclide	Half-life	Activity level, steady state (μCi/mℓ)
^{16}N	7.3 sec	100
^{17}N	4.1 sec	800
^{38}K	7.7 min	5×10^{-2}
^{41}Ar	1.8 hr	4×10^{-2}
^{18}F	1.9 hr	4×10^{-2}
^{56}Mn	2.6 hr	5×10^{-3}
^{64}Cu	12.8 hr	3×10^{-4}
^{24}Na	15 hr	1×10^{-3}
^{187}W	24 hr	3×10^{-3}
^{59}Fe	45 days	1.1×10^{-5}
^{182}Ta	111 days	0.6×10^{-5}
^{60}Co	5.3 days	2.5×10^{-5}

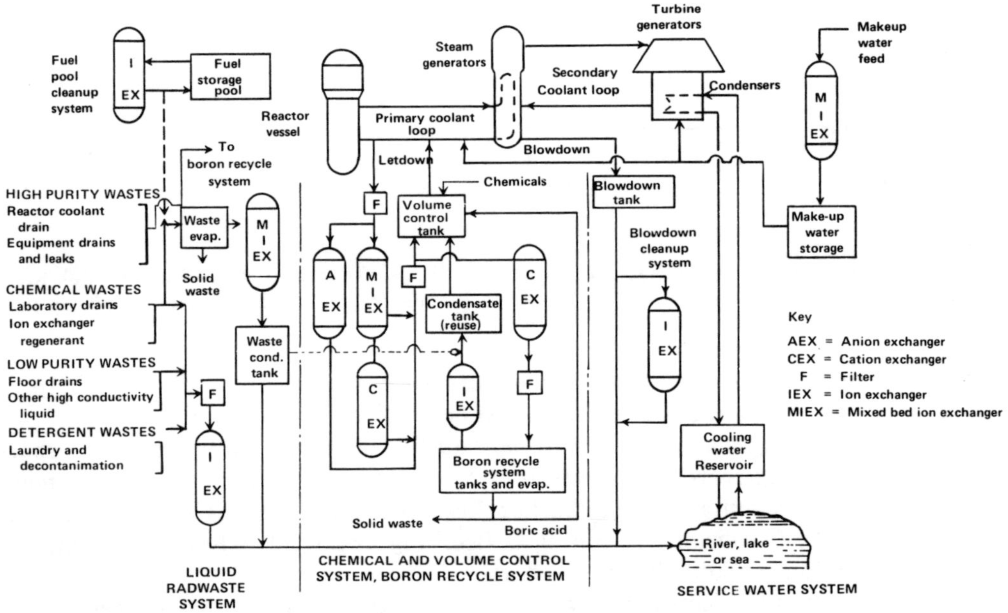

FIGURE 3. Liquid stream treatment system for typical pressurized water reactors. (From Lin, K. H., 79th Natl. Meeting, American Institute of Chemical Engineers, Houston, March 1975. With permission.)

purity, type of reactor, and neutron flux. The radioactivity level is also influenced by the ratio between the primary coolant volume and the purification flowrate. In this respect, there are differences between PWR and boiling water reactor (BWR) power plants, as indicated by Figures 3 and 4.

III. THE ROLE OF ION EXCHANGERS IN A NUCLEAR POWER PLANT

Ion-exchange materials are used extensively in the nuclear power industry for the following (Figures 3 and 4):[5-8]

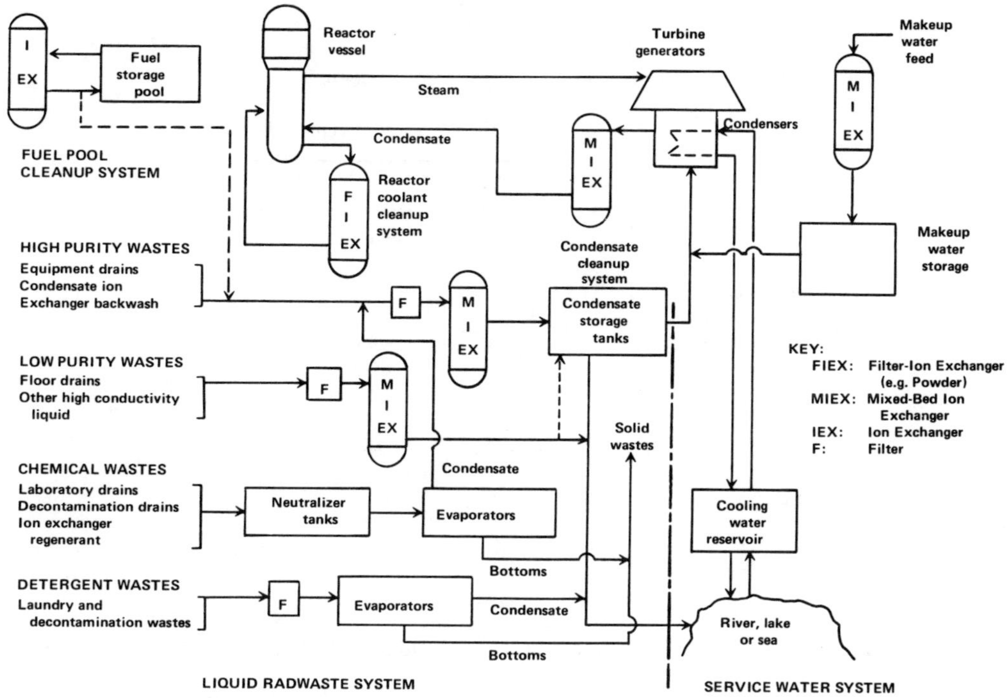

FIGURE 4. Liquid stream treatment system for typical BWR. (From Lin, K. H., 79th Natl. Meeting, American Institute of Chemical Engineers, Houston, March 1975. With permission.)

1. Preparation of make-up water
2. Treatment of condensate
3. Purification of primary coolant
4. Maintenance of spent fuel storage pools
5. Treatment of low-level radioactive wastes
   a. High-purity, low chemical content
   b. Chemical wastes, lab drains, and laundry wastes

Some typical specifications for these various applications of pure and ultrapure water are given in Tables 4 and 5.

In each case, the form of the ion-exchange materials used will depend to a great extent on the desired result. In addition, concepts concerning the equipment required to achieve an end result are continually changing in the nuclear power industry. We will try to review these changes and their relative significance in the following sections.

A. Make-Up Water

In the preparation of make-up water, it is desirable and necessary to produce a water quality as close as possible to that which is required by the circulating system under consideration.[1] When this is accomplished, then changes due to chemical excursions are minimized and smooth operation results.

In most light water nuclear plants in service, the make-up water requirements are usually satisfied with a small volume output system. This is understandable, since the circulating stream contains radioactivity and is therefore constructed with a high degree of integrity with respect to leakage. In addition, in the case of BWR plants, the steam is not withdrawn for processing. All the steam leaving the reactor is returned as

TABLE 4

Typical Reactor Water Requirements

Process stream	PWR Primary coolant	BWR Generator	BWR Condensate
Component			
H_2 (cc/kg)[a]	15—35	none	none
Dissolved O_2 (ppm)	≤0.01	≤0.01	10—30
Halogens (ppm)	≤0.1	≤0.1	≤0.05
Total dissolved salts (ppm)	≤1.0	≤1.0	≤0.05
Normal pH (77°F)	9.5—10.5	7.0—7.5	7.0—7.5

[a] At STP.

TABLE 5

Typical Auxiliary Water Requirements Common to Both PWR and BWR Systems

Component	Make-up water	Spent fuel storage	Wastewater Reuse	Wastewater Discharge
Halogens (ppm)	≤0.1	n.a.[a]	≤5	n.a.
Dissolved O_2 (ppm)	≤0.1	n.a.	n.a.	n.a.
TDS (ppm)	≤0.05	n.a.	≤0.5	n.a.
Color (APHA[b] units)	n.a.	0—5	0—5	≤10
Turbidity (APHA units)	n.a.	0—5	0—5	≤100
pH (77°F)	Footnote c	7.0—7.5	7.0—7.5	7.0—7.5
Activity level (μCi/ℓ)	n.a.	≤100	≤100	≤10
Silica (ppm as SiO_2)	<0.01	—	—	—
Conductivity	<1.0	—	—	—

[a] Not applicable.
[b] American Public Health Association.
[c] Make-up water pH is adjusted for the reactor system being designed.

treated condensate so that the circulating water inventory remains relatively constant. A make-up water treatment system which might apply is shown in Figure 5.

B. Condensate Treatment

In the usual case, the condensate will contain only very small amounts of ionic contaminants that result from steam carryover, minor amounts of corrosion products from the turbine or condenser surfaces or condenser leaks.[7] Condensate treatment therefore is necessary for the removal of these constituents so that they are not returned to the core vessel to become activated and concentrated.[9] In general the steam will carry over appreciable amounts of NH_3 (if used in the system), which is beneficial and helps reduce the corrosion rate especially at various heat-exchanger surfaces. Figures 6, 7, and 8 indicate some different points of application for condensate treatment from those shown in Figure 4.

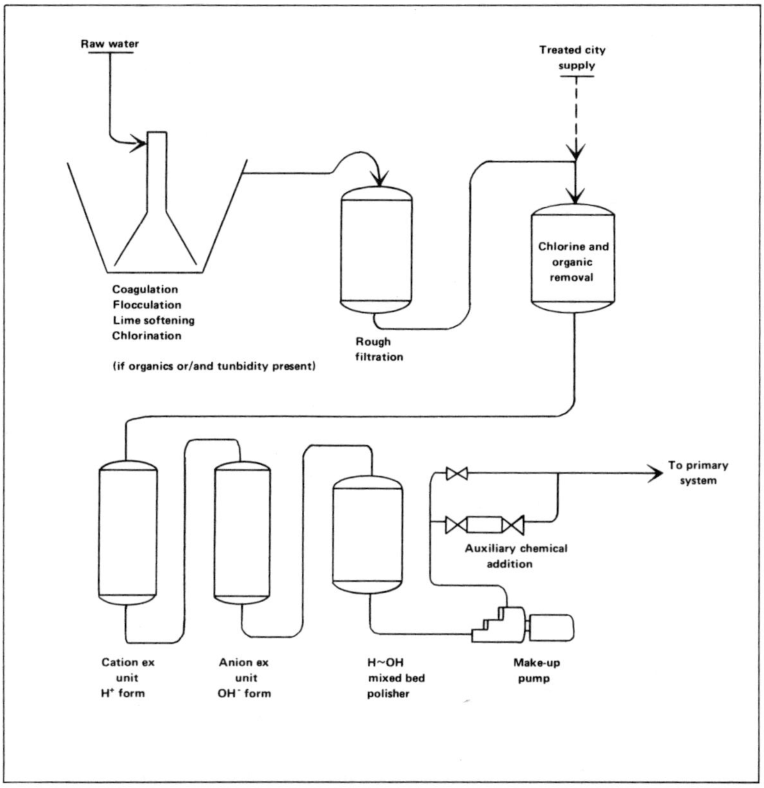

FIGURE 5. Make-up water treatment system.

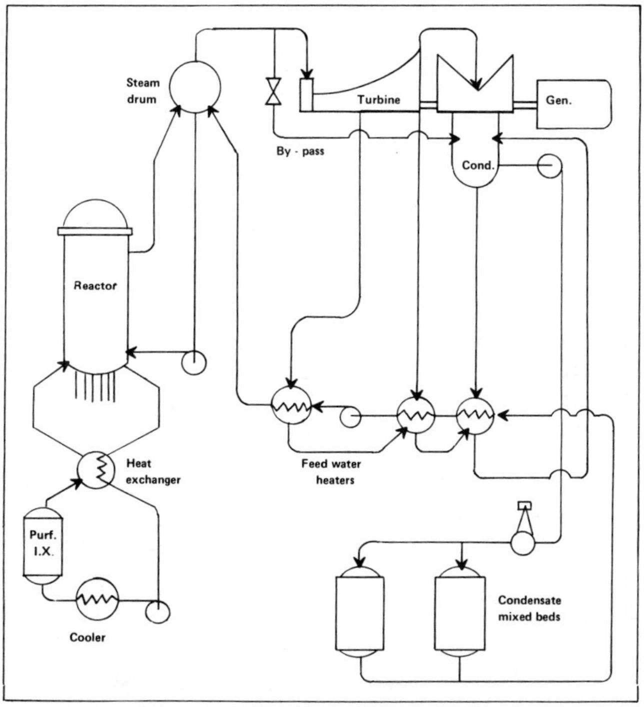

FIGURE 6. Big Rock BWR plant direct cycle system.

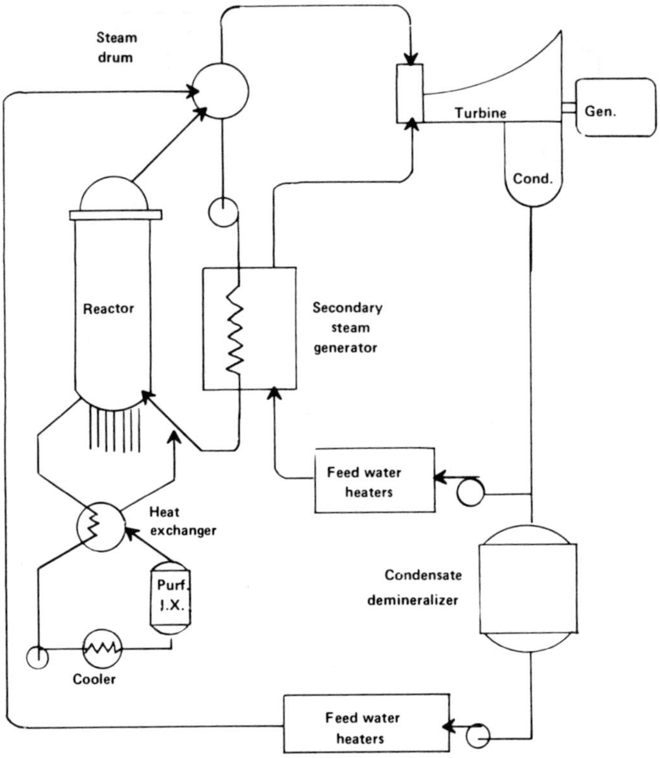

FIGURE 7. Dresden BWR plant dual cycle system.

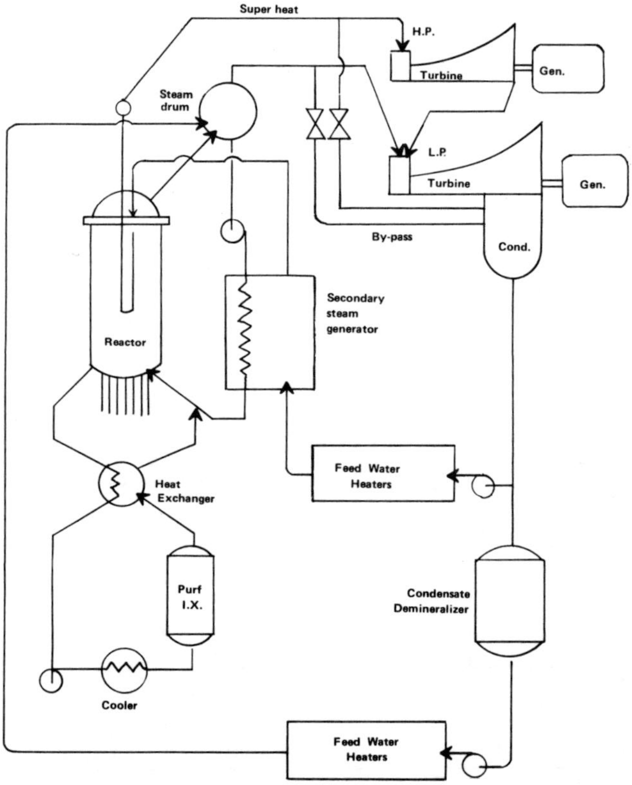

FIGURE 8. Dual cycle BWR system with superheat.

TABLE 6

Comparison of Advantages and Disadvantages of Condensate Treatment Systems

Treatment system	Advantages	Disadvantages
Conventional mixed-bed polisher	Low pressure drop	High capital cost
	Good reserve in event of condenser leak	Somewhat limited by flowrate
	Low leakage levels	
	Relatively low operating costs	
Powdered ion-exchange materials	Low capital cost	High pressure drop
	Not flowrate limited	No reserve capacity
	Good filter efficiency	High leakage rate
	Low volume of waste disposal	Relatively high operating cost
	High rate of reaction	Not good if TDS $\geqslant$ 500 ppb

In most plants, conventional ion-exchange mixed beds are used in large vessels (50 ft^3 or more), while in others, small amounts of powdered,[10] or microfine beads of ion-exchange materials are used in small equipment (Volume I, Chapter 6). There are advantages and disadvantages for each system, as shown in Table 6.

There are plants with both condensate treatment systems in use at the same time. It would appear that it will be some time before the industry is certain that one system has a technical advantage over the other.

C. Primary Coolant Purification

1. NH_4–OH Systems

Primary coolant purification is used in both the PWR and BWR systems.[11] In the PWR plant, the process is carried out as shown in Figure 3 using an appropriate H–OH or salt form mixed bed on a side-stream loop. In the BWR application, it has become common practice to take the water for coolant purification at some point near the core, since it is at this point that the highest concentration of activated corrision products exists. Also, in BWR plants, the ion-exchange mixed bed starts out in the fully regenerated (H–OH) form and is slowly converted to the ammonium (NH_4–OH) form while removing both particulate and soluble corrosion products.

An example of the performance of an NH_4–OH mixed bed in service on primary coolant for a PWR system is given in Tables 7 and 8 and shown in Figures 9 and 10. The data given here separate the exchangeable soluble corrosion products from the filterable insoluble corrosion products, often called "crud." The data indicate that an equivalent 100 full power hours utilizes only the first third of the ion exchanger for the removal of the species of interest.

Table 9 summarizes the reactor problem areas. In both systems the coolant purification system is used to reduce the coolant radioactivity level after shutdown in order to facilitate maintenance. An indication of how the coolant purification demineralizer helps to reduce the radiation levels in a PWR system is shown in Figure 11. It is evident from the data shown that the radiation levels found 15 hr after shutdown are signifi-

TABLE 7

Performance of an NH_4–OH Mixed Bed in Service on a Primary Coolant of a PWR system[11]

Radiation levels after 1700 hr full power operation (mR/hr)	Calculated for 15 hr decay (mR/hr)[a]	Actually observed (mR/hr)
9000	8901	6
6000	5934	3
3000	2967	3
1000	988	3
550	554	0.4
400	396	0.3
150	148	5
100	98.8	1.5
15	14.8	0.6

[a] Assuming that corrosion products have the following distribution after 1700 full power hours (Nuclide, % of reading): ^{60}Co, 7; ^{59}Fe, 1; ^{56}Mn, 1; ^{54}Mn, 73; ^{55}Fe, 18.

TABLE 8

Distribution of Soluble Corrosion Products on an NH_4–OH After an Equivalent of 100 Full Power Hours of Operation[11]

Sample vol as % of bed depth	Ionic conc meq/mℓ of mixed bed						
	NH_4	Fe^{2+}	Cr^{3+}	Ni^{2+}	Co^{2+}	Mn^{2+}	Total
0—25	0.21	0.18	0.175	0.015	0.007	0.003	0.590
25—52	0.41	0.12	0.048	0.010	0.005	0.003	0.596
52—75	0.50	0.066	0.019	0.008	0.004	0.003	0.600
75—100	0.54	0.035	0.012	0.006	0.003	0.004	0.600

Distribution of Insoluble Corrosion

Products (Crud) on an NH_4–OH Mixed Bed after 100 equivalent Full Power Hours of Operation

Sample vol as % of bed depth	Metal oxides (mg/mℓ of resin)
0—10	325
10—20	265
20—30	290
30—40	115
40—50	26
50—60	15
60—70	10
70—80	8
80—90	9
95—100	10

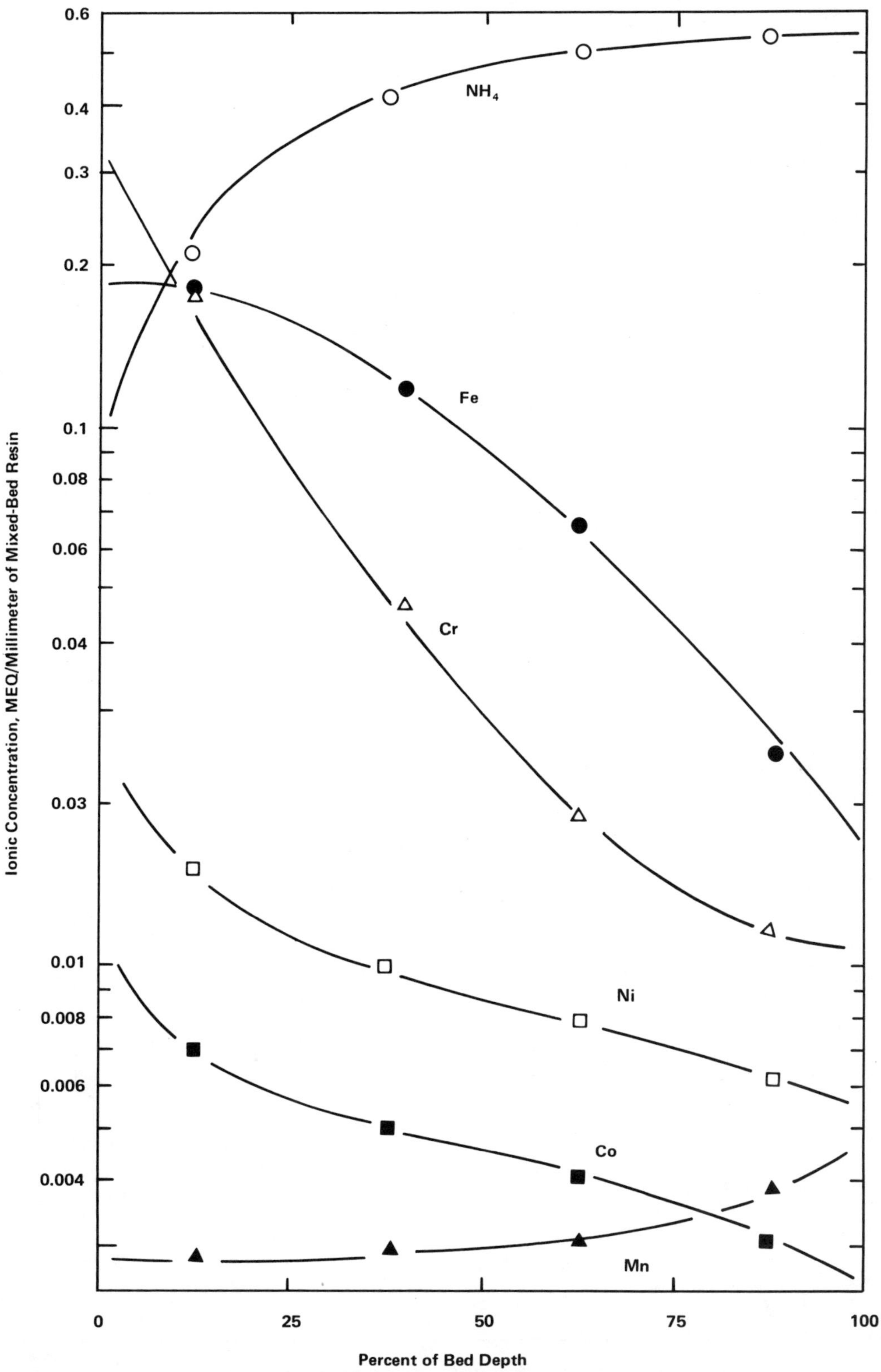

FIGURE 9. Distribution of soluble corrosion products in an NH_4–OH mixed bed. (From Simon, G. P., Abrams, C. M., and Lindsay, W. T., Jr., WAPD-TM-215, Bettis Atomic Power Laboratory, Pittsburgh, July 1960. With permission.)

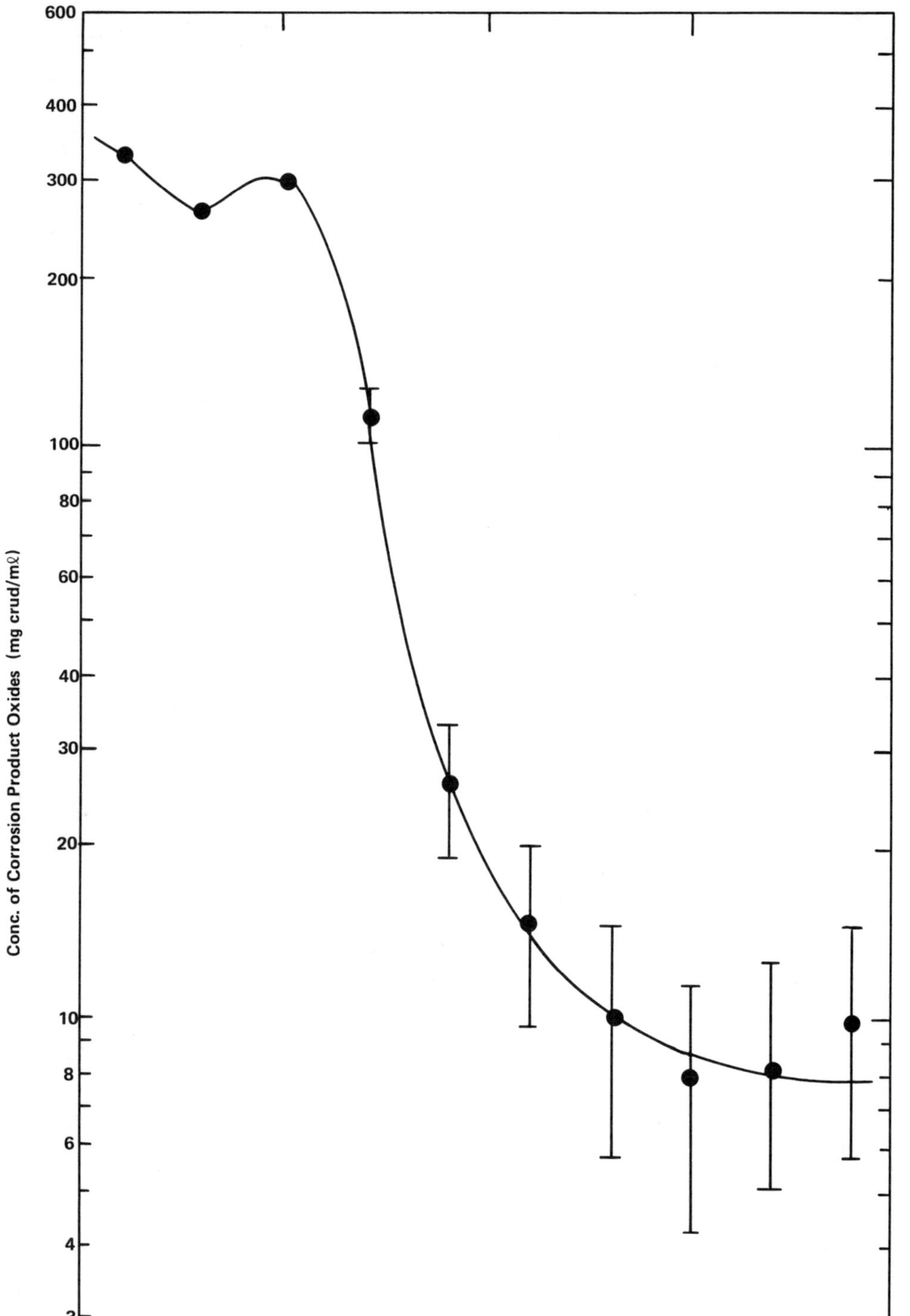

FIGURE 10. Distribution of insoluble corrosion products in an NH_4–OH mixed bed. (From Simon, G. P., Abrams, C. M., and Lindsay, W. T., Jr., WAPD-TM-215, Bettis Atomic Power Laboratory, Pittsburgh, July 1960. With permission.)

TABLE 9

Reactor Coolant Problem Areas

Type	Core	Primary system	Auxiliary system and waste disposal
PWR	1. Clad integrity 2. Clad corrosion 3. Fission product release 4. Core deposits 5. Chemical control 6. Radiolysis of coolant 7. Control-rod wear	1. Corrosion and wear 2. Activity transport 3. Contamination 4. Coolant purification 5. pH control 6. Poison control 7. Decontamination	1. Water preparation 2. Poison prep, storage and removal 3. Gaseous activity removal 4. Conc of active wastes
BWR with or without superheat	1. Clad integrity 2. Clad corrosion 3. Fission prod. release 4. Core deposits 5. Radiolysis of coolant 6. Control-rod wear	1. Corrosion and wear 2. Activity transport 3. Contamination 4. Coolant purification 5. Decontamination 6. Turbine contamination 7. Feed water purity 8. Condensate treatment	1. Water preparation 2. Gaseous activity 3. Recombination of gases 4. Conc. of active wastes

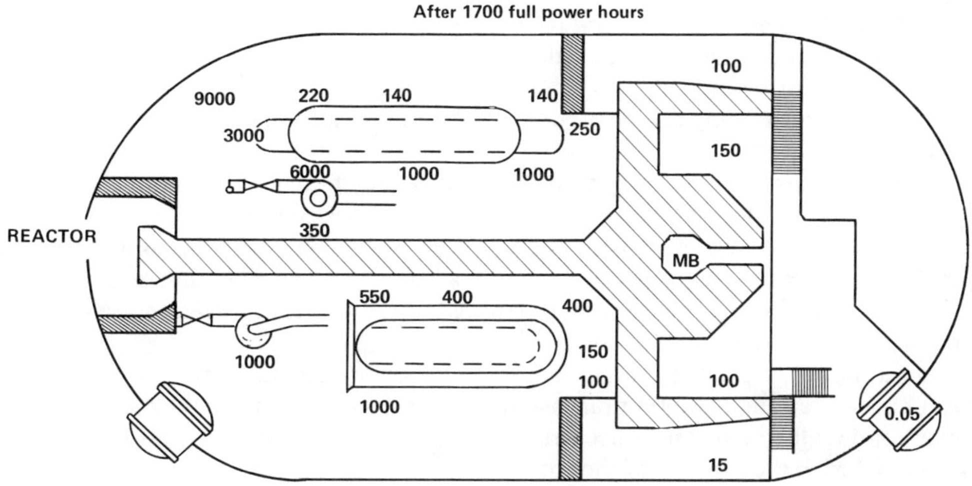

A

FIGURE 11. Shippingport BWR boiler chamber radiation levels. All readings in milliroentgens per hour. (From Petro, W. M., Bettis Tech Review WAPD-BT12, April 1959. With permission.)

cantly lower than what would be expected from normal decay, assuming some ideal distribution of ^{59}Fe, ^{60}Co, ^{58}Mn, etc. The lower radiation levels found are primarily due to the fact that the primary system NH_4–OH mixed bed was in operation during the 15 hr cooldown period after 1700 hr of full power operation.[12] It is also clearly indicated that radiation levels near the operating mixed beds do not decrease to the same extent for the 15 hr period of interest.

2. Deborating System

In PWR systems, it has become common practice to use boron as a means of con-

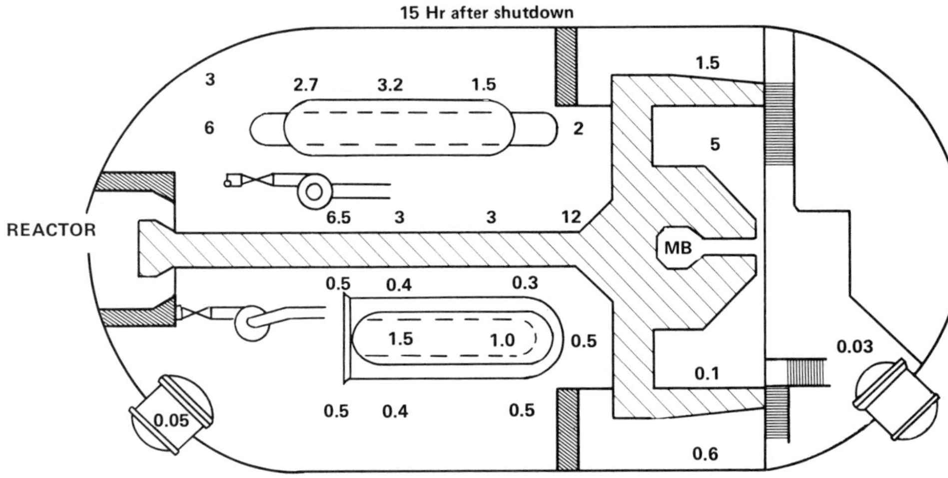

FIGURE 11 B

trolling core activity and local variations in core reactivity. This effect has been called boric acid "shim" control, and as the term implies, it involves very small changes in neutron flow levels. The use of boron in PWR coolants brings about the following advantages:

1. Fewer and less expensive central rods can be used
2. Simplified core design can be used.
3. A more uniform power distribution is achieved in the core.
4. A greater fuel loading can be used to achieve an extended core life.

Control rods are still necessary to keep up with and control those rapid changes due to power demands. Boron "shim" control, however, is used to advantage during start-up when the coolant water density is decreasing as the system goes from ambient to operating temperature. During core life, the amount of samarium and xenon poisons in the fuel increases, and proportional decreases of boron concentration are necessary to maintain uniform operations.

Deborating anion-exchange materials are used during start-up to reduce initially high boron concentrations (≈ 3000 ppm as B) to operating levels (≈ 1800 ppm as B) and reduce continually the boron concentration as the fuel depletion and poisoning changes take place during the lifetime of the core.

The removal of borated "shim" from the primary coolant system of land-based nuclear power plants must be carefully controlled. Large quantities of boron, with or without radioactivity, are undesirable contaminants for the environment. Therefore, it has become general practice to treat any and all solutions containing boron for final complete and safe disposal. As indicated in Figure 6, this is accomplished by using a deborating strong-base, type I anion exchanger and recycling the boron-free water back to the primary coolant systems. In addition, the ion exchanger regenerants containing boron and the more concentrated boron solutions encountered during start-up are evaporated to provide a concentrate for final disposal.

3. BWR Entrainment from Steam Generator

At low pressures and moderate temperatures, solute volatility is a minor source of transport compared to the physical carryover or entrainment.

The potential, therefore, exists in the superheater for accelerated attack by chloride solutions; it is also important to maintain chloride ion concentrations at low levels. To prevent the development of concentrated solutions, it would be necessary to maintain all surfaces at 590°F at all times. Then droplets carried in the steam without being deposited would dry out and the salt volatilize into the steam whenever the bulk concentration is less than the minimum solubility for the conditions in the superheater.

Caustic solutions, with the exception of LiOH, will not dry out. These solutions are aggressive to almost any ordinary material of construction used in the superheater. Therefore, free caustic in the steam generator source should be avoided at all times.

4. Turbine Deposits

In high-pressure and supercritical cycles in nuclear systems, problems arising from the transport of volatile solutes will be further complicated by the radioactivity of these volatile solutes. It is essential that this aspect of system operation be thoroughly reviewed before large-scale nuclear plants of this type are built. The transport and retention of iron, for example, will probably be less than 0.1 ppb. For a 1000 MWe plant, the steam flow will be of the order of 10 million lb/hr, or 80×10^9 pounds per year. The iron transport will therefore be no less than 0.8 lb/year. This, together with other material, will be radioactive and will be a very real nuisance during turbine maintenance or repair.

D. Spent Fuel Storage Pools

The primary purpose of the ion-exchange materials used is to provide an optically clear water. This is necessary, since all the manipulations required for packaging spent fuel elements must be carried out underwater to keep radiation hazards to a minimum. In special cases, the H−OH form mixed bed is used to remove specific radioactive nuclides which find their way into the pool water from defective fuel elements and cladding failures. Relatively high concentrations of radionuclides may also occur in these facilities from mechanical operations carried out in the pool.

E. Waste Treatment

There are two techniques for recovering reactor wastewater.[13] In the first case, the wastewater is filtered, stored for the reduction of short-lived nuclides, and eventually demineralized and recycled. In order to provide the maximum decontamination factors, it has become common practice to employ H−OH form mixed beds in this application. In the second case, the wastewater is evaporated; normally, the evaporator condensate is polished with an H−OH mixed bed, and this treated water is also recycled back to the reactor make-up system.

Some typical decontamination factors that can be achieved with different ion-exchange treatment systems are shown in Table 10.[5]

The heaviest load for the waste disposal system occurs during plant shutdown for refueling. A typical rad waste system is shown in Figure 12. As indicated, plant letdown comes to the rad waste system through the primary purification demineralizer, and after flashing for the removal of radioactive gases, it is stored in one of three available large liquid waste hold-up tanks. Other hold-up storage facilities are available for reactor system drains, radiochemistry laboratory drains, decontamination drains, and building sumps. Evaporator bottoms and unrecoverable wastes are concentrated for final disposal.

In BWR systems, the full-flow condensate demineralizers are the main source of radioactive wastewater. A small amount of radioactivity is entrained irreversibly in the ion exchanger, but the bulk of the radioactivity is found in the spent regenerant chemicals or in the demineralizer backwash water.

TABLE 10

Decontamination Factors Obtained Using Various Ion-exchange Systems[5]

Reactor system	Treatment system and source	Decontamination factor (average for system lifetime) I	Cs,Rb	All other solubles
PWR	Mixed bed ($L^+-BO_3^{-3}$ form)	10	2	10
	Cation bed (H^+ form)	1	10	100 (cations)
	Mixed bed (H^+-OH^- form)	100	2	100
	Anion bed (OH^- form)	100	2	100 (anions)
BWR	Mixed bed (H^+-OH^- form)			
	Primary coolant	10	10	10
	Steam condensate	1000	10	1000
	Clean waste	100	10	100
	Dirty waste	100	2	100
	Cation bed (H^+ form) Dirty waste	1	10	100 (cations)
	Powdex in any system	10	1	10

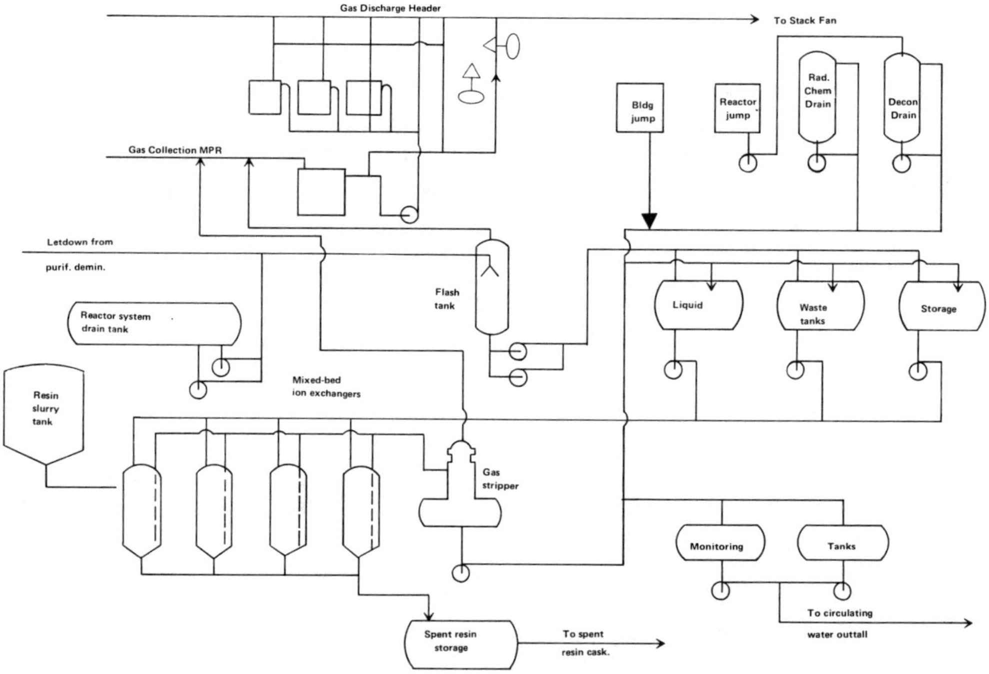

FIGURE 12. Typical rad waste system. (From Baumann, E. W., 147th Meeting Am. Chem. Soc., Philadelphia, 1964. With permission.)

Except for materials with a high turbidity, equipment drainage and backwash water from condensate polishers are usually recoverable by filtration followed by ion exchange. This recovered wastewater is recycled back to condensate storage.

Regenerant chemicals are neutralized and evaporated. Solids or bottoms are sent to packaging for eventual waste shipment. In isolated cases, evaporator bottoms have been discharged into the station circulating water which goes back to the environment.

With PWR plants, the primary means of handling waste streams is by evaporation. Condensate from the evaporator is recycled to the make-up system. Evaporator bottoms are combined with other plant waste solids and packaged for eventual shipping to a suitable disposal site.

F. Gaseous Products

Gaseous products from the flash tank are collected and stored to take advantage of the natural decay. After cryogenic purification, the gases are released in the stack.

Usually all holding tanks are duplicated so that a full tank of waste can be processed through two waste treatment ion exchangers in series and the effluent returned to an empty clean hold-up tank. Whenever the contents of the holding tank meet the minimum release specifications for the location, it is discharged through a fresh ion-exchanger bed to a final gas stripper and to monitoring tanks. From the monitoring tanks the processed waste is passed to the circulating water outfall where it is further diluted before reaching the environment. Whenever the content of a holding tank does not meet the minimum release requirements after the first ion-exchange treatment, it is reprocessed through a fresh mixed bed and returned to a clean liquid waste holding tank.

The rad waste ion-exchange treatment system usually consists of four 25-ft^3 mixed beds. The mixed beds consist of a strong-acid cation exchanger in the hydrogen form and an acrylic weak-base anion exchanger in the free-base form. The use of the weak-base insures the efficient removal of radioactive iodine. While rad waste deionizers are designed for flow rates up to 50 gal/min, they are usually found to be more efficient at flow rates of 15 to 20 gal/min.

It is common practice to pass the waste through the oldest mixed bed first so that this unit can be used as a filter for the removal of crud and particulate material. This process sequence protects the newly installed ion exchangers from becoming prematurely fouled.

Experience indicates that rad waste ion exchangers are required to process about 100,000 gal of contaminated water each month. Replacement appears to average about eight (25 ft^3) charges per year. This indicates that each charge will handle about 150,000 gal of waste.

During that period when spent fuel shipments are made, there is an additional load on the rad waste system, since all the spent fuel cask decontamination solutions and flush water must be returned to a holding tank for processing.

A well-designed and properly operated rad waste system will normally release solutions to the environment which are less than the normal background radiation of the stream or river. Radiation from combined iodine and cesium radioisotopes will often be less than 10^{-8} microcuries/cc at the circulating water outfall.

IV. BREAKDOWN PRODUCTS AND ION EXCHANGERS

Studies covering radiolysis of ion-exchange resins as used in purification systems of reactors have been reported.[14,15] Radioactive species, deposited from the coolant on ion-exchange resins, can cause decross-linking or can cause the release of soluble sulfonates or amines. These, on entering the high radioactive environment of the reactor, can also become radioactive. If these entities have ionic groups, they will be picked up by the ion exchangers. Proof of the existence of these by-products from ion-exchange resins is the fact that sulfur 35 and phosphorus 32 have been found in the reactor. The latter is due to the transformation of the decay of radioactive sulfur into phosphorus 32. CO_2

was also found, believed to be due to the oxidation of organic matter of mixed beds by radioactive deposits on the resins. The cation exchangers picked up 90% of the above radioactive species.

High radioactive dosages will cause damage to ion-exchange resins as evidenced by swelling and reduction in volume capacity. Although efficiency of the exchangers is reduced, the release of sulfur, nitrogen, and carbon species into the reactor is eliminated.

Most of the radioactive nuclides found in the coolant are removed by the cation exchanger as the species are mostly cations. Therefore, some plants use a cation exchanger followed by a mixed bed containing a much higher ratio of cation exchanger to anion exchanger.[16] More complete studies on the effect of radiation on ion exchangers can be found in Volume I, Chapter 5.

REFERENCES

1. **Cohen, P.,** *Water Coolant Technology in Power Reactors,* Gordon & Breach, New York, 1969.
2. Radiological Health Handbook, Environmental Protection Agency Sanitary Engineering Center, U.S. Department of Health, Education and Welfare, Cincinnati, 1954, 83.
3. Sullivan, W. H., Trilinear Chart of Nuclides, Oak Ridge National Laboratory, U. S. Government Printing Office, Washington, D. C., 1957, 4.
4. **Welinsky, I. H., Cohen, P., and Seamon, J. M.,** *Proc. Am. Power Conf.,* 18, 567, 1956.
5. **Lin, K. H.,** Performance of Ion Exchange Systems in Nuclear Power Plants, in 79th Natl. Meeting, American Institute of Chemical Engineers, Houston, March, 1975.
6. **Martucci, J. A.,** Nuclear Technology, 10th Annual Liberty Bell Corrosion Course, Philadelphia, September, 1974.
7. **Moore, C. S. and Sloan, W. J.,** Very High Chemical Efficiency for Mixed Bed Ion Exchange, in 33rd Int. Water Conf., Pittsburgh, October 1972.
8. **Calmon, C. and Gold, H.,** Treatment of Industrial Wastes by Ion Exchange, in 2nd Natl. Conf. Complete Water Reuse, Chicago, May 1975.
9. **Zukerman, I. I.,** Water Technology of Nuclear Reactors, in Liberty Bell Corrosion Course, Philadelphia, September 1967.
10. **Ryan, L. F. and Brown, R. M.,** Powdered Ion Exchange in Nuclear Cycles, in 30th Annual Meeting Power Conf., Chicago, April 1968.
11. **Simon, G. P., Abrams, C. M., and Lindsay, W. T., Jr.,** The Performance of Base-form Ion Exchangers for pH Control and Removal of Radioisotopes from a Pressurized Water Reactor System, WAPD-TM-215, Bettis Atomic Power Laboratory, Pittsburgh, July 1960.
12. **Petro, W. M.,** Radiation Levels at Shippingport, Bettis Tech. Review WAPD-BT12, April 1959.
13. **O'Mara, R. L.,** Disposal of Radioactive Water from Light Water Reactors, Proc. 31st Int. Water Conf., Pittsburgh, 1970.
14. **Baumann, E. W.,** 147th Meeting Am. Chem. Soc., Philadelphia, 1964.
15. **Smith, L. L. and Groh, H. J.,** The Effect of Gamma Radiation on Ion Exchange Resins, Atomic Energy Committee Research Development, Dept. DP-549.
16. **Muller, A. C.,** *Ind. Eng. Chem.,* 5, 1254, 1959.

Chapter 6

RECLAMATION OF HIGH-SULFATE IRRIGATION DRAINAGE WATER FOR POWER PLANT COOLING

H. H. Sephton and G. Klein

TABLE OF CONTENTS

I. INTRODUCTION

Fresh water for cooling power plants planned for the California Central Valley is anticipated to rise in price and decline in availability sufficiently to justify serious consideration of irrigation tile drainage (or agricultural waste water) for this service. The latter is anticipated to increase in volume to about 500,000 acre-feet/year, containing about 3000 ppm of total dissolved solids (TDS) by the year 2000, and to reach 700 million gal/day at peak summertime flow. The main objective of this work was to develop a procedure whereby this waste water could be used in power plant cooling towers, after reducing its high scale-forming potential by softening it in an economically acceptable way.[1] A procedure was formulated utilizing a novel process sequence based on earlier developments in the Sea Water Conversion Laboratory, University of Cal.[2-4] The flow sequence proposed is shown in Figure 1, which incorporates the following consecutive steps:

1. Pretreatment of the waste water by passing it through a cation exchanger in the sodium form to reduce its scale-forming tendencies related to its calcium and magnesium content
2. Evaporation of the softened waste water by a factor of 5 to 15 in a cooling tower circuit under typical recycled coolant process conditions, with continuous addition of softened feed and blowdown of brine at controlled salinity

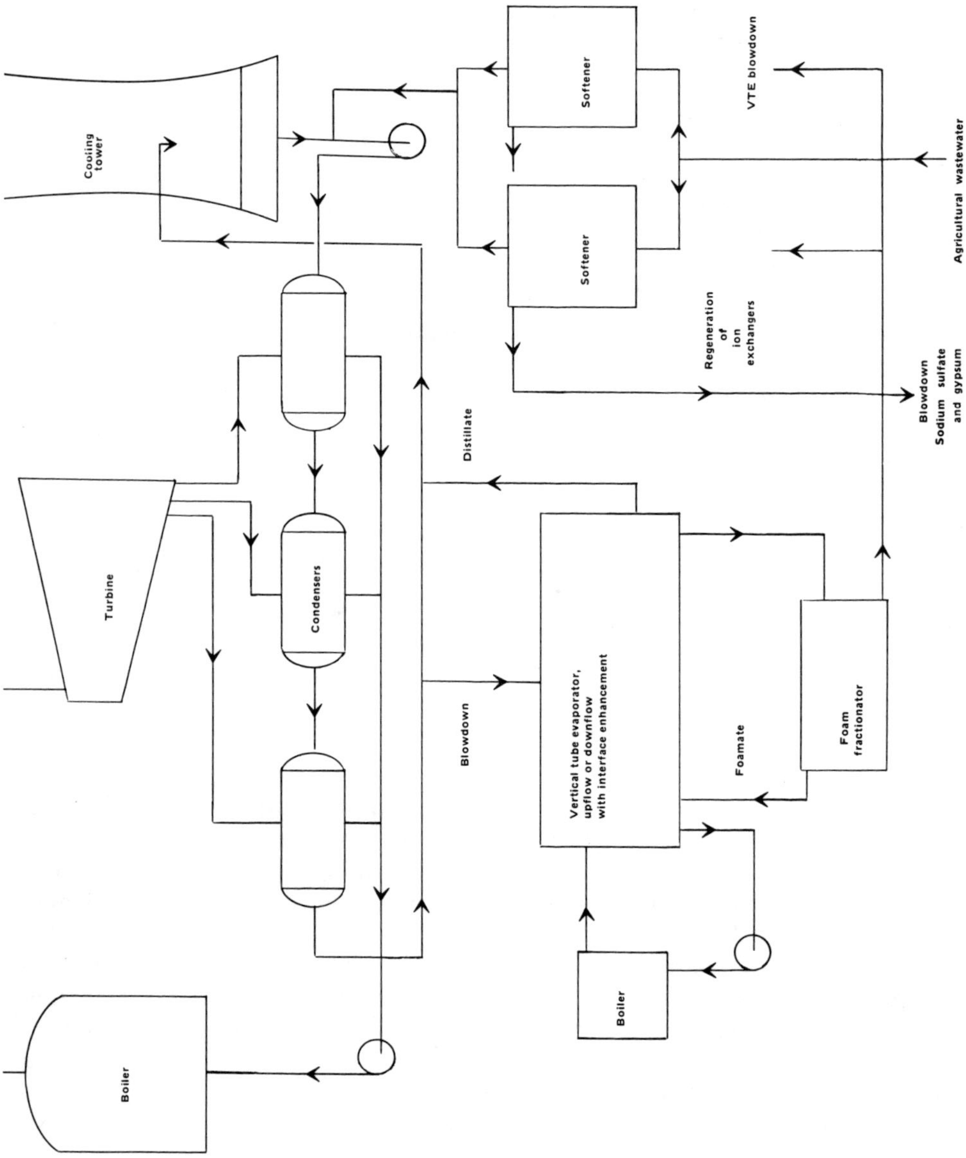

FIGURE 1. Proposed power plant coolant cycle.

3. Concentration of the cooling tower blowdown by interface-enhanced vertical tube foam evaporation (VTFE)[3] to a salinity determined by its use as a sole regenerant of the ion-exchange resin
4. Clarification to the VTFE concentrate by foam fractionation-flotation[4] as needed to provide a clear brine for regeneration of the resin and to recover the foaming agents for recycling
5. Regeneration of the ion-exchange resin, used in the first (softening) step, by a fluidized-bed procedure which separates calcium sulfate precipitate formed during regeneration from the resin bed[5]

Of primary significance in this approach was the adaptation of the ion-exchange softening pretreatment to a challenging feed water. This step, especially the prospect of regeneration solely with a blowdown concentrate, was seen as the crucial step determining the overall feasibility and economy of the process sequence. A similar pretreatment process has been envisaged for desalination by reverse osmosis by Haugseth and Beitelshees[6] and its further development described by Taylor and Haugseth.[7] Their results and our own appear to be mutually supportive.

It was anticipated that this method of softening pretreatment and brine concentration should be more economical than conventional lime- and soda-ash treatments, especially if waste heat could be used for the VTFE concentration step.[8] These methods of treating power plant coolant and blowdown could find wide application if present field tests continue to be successful. Under EPA guidelines, desalting and recycling of cooling tower blowdown has to be considered even when fresh water is used as coolant. Also, softening treatment is frequently necessary in order to permit high factors of concentration in the interest of water economy and to limit the volume of blowdown to be disposed of. In addition to possibly turning an environmental liability into an asset for the irrigation drainage case, this approach could have wider applications, such as in the utilization of municipal and industrial waste waters and of brackish water in cooling towers and for seawater desalination.

II. ION-EXCHANGE SOFTENING AND REGENERATION

A. Relation of Raw Water Composition to Regeneration Mode

In earlier work at the University of California Sea Water Conversion Laboratory, cation exchange with strong-acid resins was considered for seawater softening prior to evaporation.[3,9-13] In this fixed-bed process, evaporator blowdown was used as the sole regenerant, so that all costs associated with the purchase and preparation of regenerant chemicals were eliminated. For concentration factors greater than 4, difficulty was encountered with precipitation of calcium sulfate during fixed-bed regeneration.[10]

The compositions of typical sea water and agricultural waste water of the San Joaquin Valley type are shown in Table 1. The TDS concentration of this agricultural waste water varies seasonally between about 2000 ppm and 7000 ppm, but the relative concentrations of the various constituents are fairly constant. The salient differences with respect to sea water are seen to be the considerably higher relative concentrations of calcium and sulfate, and the lower TDS concentration in agricultural waste water. These factors, at comparable regenerant concentrations, may be shown to contribute to a greater tendency for calcium sulfate precipitation to occur during fixed-bed regeneration with reject brine from agricultural waste water than with sea water.

For conditions of practical interest, this tendency has been confirmed by laboratory experiments, which indicated that regeneration in the fixed-bed mode was not feasible because of calcium sulfate precipitation in the exchanger bed. When sulfate ions were left out of the regenerant formulation to eliminate this difficulty, the amount of rege-

TABLE 1

Typical Compositions of Sea Water and Agricultural Waste Water

	Sea Water[a]			Waste water[b]		
Ion	ppm	Normality	Fraction of normality	ppm[c]	Normality	Fraction of normality
Na^+ (+ K^+)	11,400	0.49	0.79	700	0.030	0.60
Mg^{++}	1,300	0.11	0.18	100	0.010	0.20
Ca^{++}	420	0.021	0.034	200	0.010	0.20
Cl^- (+ NO_3^-)	19,800	0.56	0.90	350	0.010	0.20
SO_4^{--}	2,800	0.06	0.10	1,700	0.035	0.70
HCO_3^-	150	0.002	—	300	0.005	0.10
Total	35,870	0.62	1.00	3,350	0.050	1.00
				(2,000—7,000)[d]	(0.03—0.10)[d]	

[a] Values rounded off from those given in Reference 14.
[b] Based on chemical analyses of samples collected at Firebaugh, California. Underlying data provided by California Department of Water Resources.
[c] Representative mean values.
[d] Approximate limits of seasonal variation.

nerant that could be expected to result from concentration of the softened water proved insufficient to sustain continued cyclic operation.[5,15] The sea water softening process with fixed-bed counterflow regeneration thus did not appear applicable to the present problem.

B. Waste Water Softening

Removal of calcium and magnesium from the agricultural waste water is achieved with conventional downflow through a fixed bed of a standard polystyrene divinyl benzene sulfonate resin in the sodium form. The softened water, primarily a solution of sodium sulfate and chloride, after concentration in cooling tower and evaporator, then serves as the sole regenerant.

The formation of calcium sulfate during regeneration is an example of ion exchange accompanied by precipitation. In the sulfate removal process of George et al.,[16] barium sulfate was formed. In their work, a series of stirred tanks in cascade arrangement was employed to separate the precipitate from the resin. In the deionization process proposed by Popper et al.,[17] a calcium hydroxide precipitate is allowed to form in a fixed- or pulsed-bed, where it moves with the velocity of the fluid phase because of the sparseness of the crystals and their nonadherence to the resin particles.

In the regeneration of the exchanger used to soften the agricultural waste water under consideration, a copious amount of calcium sulfate precipitate tends to form. In the successful method adopted here, regeneration is therefore carried out in the upflow direction, at a rate sufficient to fluidize the bed and to carry calcium sulfate particles precipitated during regeneration out of the column through openings in its upper part. At the same time, the flowrate must be low enough to prevent carryover of resin particles with the effluent. When a flowrate satisfying these requirements was used, however, it was found that a single-pass application of the regenerant brine produced per cycle did not regenerate the resin adequately. Regenerant was therefore recycled in the manner shown in Figure 2. A volume of spent regenerant, equal to several volumes of evaporator-brine concentrate corresponding to one softening cycle, was collected in a spent-regenerant storage tank during prior operation. Formation of

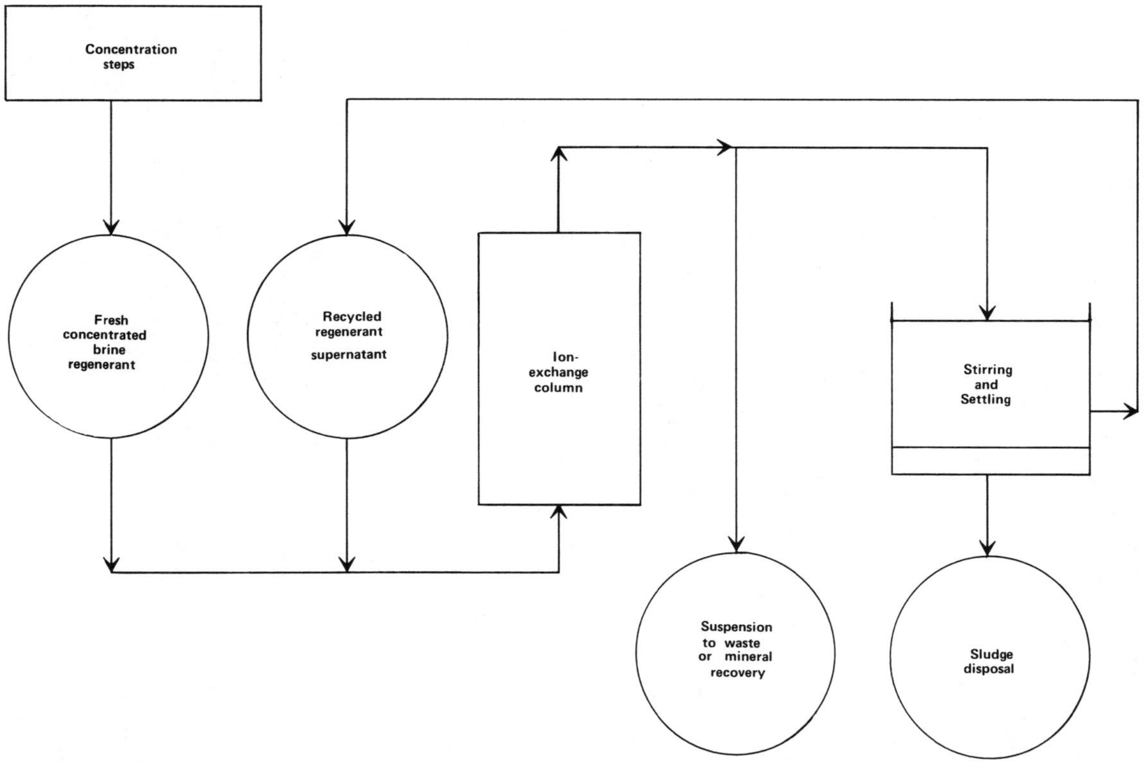

FIGURE 2. Fluidized-bed regeneration with waste brine. Regenerant used in previous cycles, from which calcium sulfate precipitate has been removed by stirring and settling, is applied first. Fresh VTFE brine produced in the concentration steps is used as a polisher.

larger calcium sulfate crystals was hastened by stirring, and after separation of this brine from the crystals by settling, was pumped into a spent-regenerant feed tank and from there passed upflow through the exchanger bed at a rate high enough to rapidly induce fluidization. When the desired bed expansion was reached, the flowrate was reduced to keep the upper level of the resin bed stationary.

The first portion of concentrated regeneration effluent, equal in volume to that of the evaporator blowdown produced per cycle, is wasted. After the appropriate volume of the spent-regenerant solution has been applied, the amount of evaporator blowdown brine equivalent to one softening cycle is sent through the column to complete the regeneration step with the solution containing the least quantity of calcium and magnesium ions available for this purpose.

C. Process Development

Steps required to carry the regeneration process from the conceptual to the pilot plant stage have included:

1. Selection of a suitable exchanger material
2. Search for combinations of bed depth, flowrate, and free-board suitable for separation of the precipitate from the fluidized resin during regeneration
3. Delimitation of the range of suitable total regenerant concentrations
4. Development of a simple and effective method to separate the precipitate from the regeneration effluent
5. Establishment of the amount of regenerant brine to be recycled
6. Scheduling of the application of various brine fractions
7. Demonstration of the cyclic operability of the process using solely the amount of fresh reject brine generated
8. Measurement of the effective exchange capacities of the resin for calcium and magnesium
9. Design of a pilot plant unit
10. Testing of this unit in multicyclic operation with both simulated and actual agricultural waste water

Laboratory experiments designed to implement Steps 1 through 8 have been described in detail by Jarvis[15] and Klein et al.[5] Among the salient findings were that (1) stirring and settling provided suitable separation of calcium sulfate crystals from regenerant brine to be recycled, (2) about 130 gal of recycled brine per cubic foot of resin, followed by the amount of evaporator reject brine corresponding to that generated during one ion-exchange cycle, were able to regenerate the resin and produce an effective exchange capacity for calcium plus magnesium of the order of 80% of the total theoretical capacity of 1.9 eq/ℓ of resin, and (3) a regeneration flowrate of the order of 10 gpm/ft^2 permitted sustained cyclic operation without a systematic reduction in effective capacity. This last finding is illustrated in Table 2 for synthetic raw agricultural waste water having the relative concentrations shown in the typical data of Table 1, and for regenerant brines reflecting concentration factors of 10 and 20, respectively. The column diameter was 2 in. with bed depths of 36 and 30 in. for the respective concentration factors. To permit direct comparison of the results corresponding to the two concentration factors, the amount of product produced per cycle, shown in Table 2, was expressed in terms of equivalents of product solution on the basis of 1 eq of resin. The exhaustion step of each cycle was stopped when the calcium concentration in the column effluent had risen to 10% of the influent level. For a concentration factor of 20, the average calcium concentration in the softened product

TABLE 2

Effect of Concentration Factor n on Amount of Product in Cyclic Fluidized-bed Operation[a]

	Equivalents of product per equivalent of resin	
Cycle no.	(n = 20)	(n = 10)
1	2.11	1.88
2	2.13	1.84
3	2.12	1.88
4	2.09	1.87
5	2.11	1.86

[a] Raw water normality = 0.1.

water amounted to only about 3% of the raw water value, corresponding to 97% calcium removal and an effective capacity for calcium amounting to 41% of the total theoretical exchange capacity of the resin. Magnesium removal was 94%, and the effective capacity for magnesium, 40% of the total. Utilization of the resin for hardness ions was thus 81% of the total capacity.

D. Pilot Plant Design and Operation

Based on the laboratory experiments, a pilot plant was designed capable of softening about 2000 gal of the most concentrated waste water to be expected in a two-cycle, 8-hr day. A schematic flowsheet is shown in Figure 3.

The ion-exchange column, constructed from an acrylic cylinder, had an effective cross-sectional area of 1 ft^2 and was 8 ft high. A stainless-steel screen near the bottom, positioned over a perforated plastic plate, prevented resin loss at the lower end and supported the exchanger bed. Overflow outlets 7 ft above this screen were provided. An overall view of the column is shown in Figure 4.

Duolite® C-20 (Diamond Shamrock Corporation) cation-exchange resin was arbitrarily selected as a standard product representative of its general type. The theoretical capacity of this resin in the sodium form is 2.1 g-eq/ℓ.

After some preliminary cyclic runs with synthetic simulated agricultural waste water had demonstrated satisfactory operability and production volumes as expected, actual agricultural waste water collected by subterranean drain tiles at Firebaugh, California, and provided by the California Department of Water Resources, was used. The respective TDS levels of these waters were 6700 and 3600 ppm. The softening flowrate was 5 gpm/$ft.^2$

During regeneration, the bed, 3 ft deep in the fixed-bed condition, expanded to a fluidized height of 6 ft at a flowrate of about 10 gpm/$ft.^2$ Precipitation of calcium sulfate in the column, or in the regeneration effluent, was observed.

To ensure adequate regeneration and removal of the precipitate, a 400-gal inventory of previously used regenerant was applied first. This was followed by a fresh batch of concentrated soft waste water (blowdown). The first portion of the regeneration effluent, equal in volume to this fresh batch, was discarded in order to keep the volume of the regenerant to be recycled constant. The remainder of the effluent was reserved for reuse. On this basis, sustained cyclic operation of this resin column was found to be achievable with an effective resin capacity for calcium above 40% of its total equi-

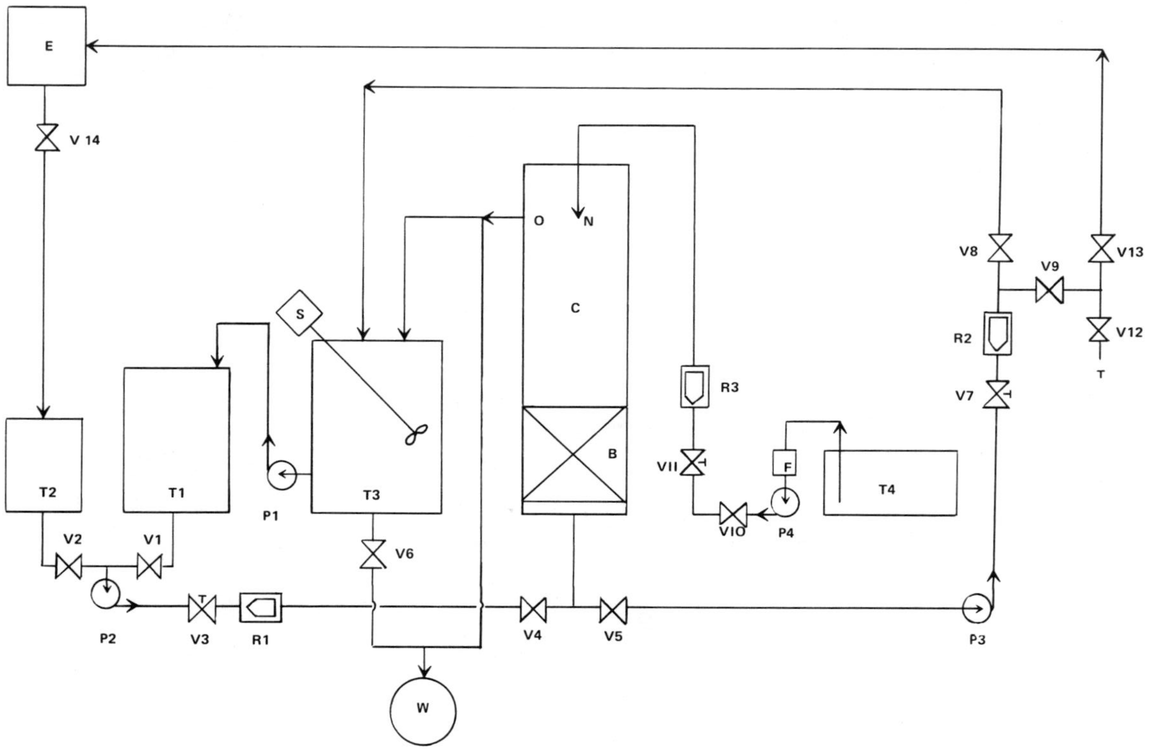

FIGURE 3. Schematic diagram of ion-exchange pilot plant. B, resin bed; C, ion-exchange column; E, evaporation-concentration steps; F, filter; N, nozzle; O, overflow openings; P, pumps; R, rotameters; S, stirrer; T1, etc., tanks; T, sampling tap; and V, valves.

FIGURE 4. Pilot plant using irrigation drainage as power plant coolant. Shown are the cooling tower, VTFE, and ion-exchange equipment.

librium capacity when the regenerant TDS level was ca. 70,000 ppm. An approximately 20-fold concentration of the original waste water after softening was thus needed to satisfy the conditions for regeneration.

The production volumes for the first ten cycles run with the pilot-plant column are shown in Table 3. In the first four cycles, simulated agricultural waste water was softened. In the remaining cycles, actual agricultural waste water trucked in from Firebaugh was used.

The production volumes for the first four cycles are seen to be quite constant. They correspond closely to the volume calculated on the basis of the data obtained earlier with the laboratory columns. The variation observed in the production volumes obtained in the runs with actual agricultural waste water is attributed to fluctuations in the raw water composition. In the absence of a trend, there is no indication of a systematic reduction of exchange capacity from cycle to cycle. As before, the resin removed about 97% of the calcium, and 94% of the magnesium from the raw water.

TABLE 3

Cyclic Performance of Ion-Exchange Pilot-Plant Column

Cycle No.	Product volume (gal/3 ft^3 resin)
1	1023
2	980
3	986
4	996
5	1170
6	1174
7	871
8	1246
9	1237
10	1116

III. COOLING TOWER AND BRINE CONCENTRATION FACILITIES

The proposed power plant coolant flow diagram shown in Figure 1 applies ion-exchange softening to the entire coolant makeup stream, rather than using the conventional side-stream softening. The softened makeup is then preconcentrated to a constant, maximal permissible dissolved solids content consistent with an acceptable heat rejection performance in the power plant condenser and cooling tower loop, with addition of softened makeup and rejection of coolant blowdown from this loop. This coolant blowdown is then further concentrated by VTFE to provide a blowdown of sufficient concentration to serve as the sole regenerant for the ion-exchange resin.

A. Cooling Tower Design and Operation

The 35-ft tall cooling tower, shown in the left background in Figure 4 was designed to simulate the significant process conditions pertaining to a large power plant installation. It had a packed-deck section of 20 ft with a flow cross-section of 2 $ft.^2$ Two types of wet-deck materials were used: standard corrugated sheets of melamine-impregnated neoprene-asbestos and self-extinguishing polyvinyl chloride, removably installed in a vertical orientation and spaced 1 in. horizontally. Eight sections were used, four of each type of material; the wet-deck sections were 16 in. deep, with 12-in. vertical spacings between sections. This coolant cascade channel was made of marine-grade plywood and was supported by an angle iron frame of 7 ft × 7 ft, 33 ft high. Coolant was continuously recirculated between this tower and a 40-kW heat source and sprayed over the top wet-deck section through four stainless-steel spray-cone nozzles, at rates of flow variable between 2 and 10 gpm/$ft.^2$ Air flow upward through the wet-deck sections was by induced draft, using a commercial cooling tower fan, continuously belt-driven by an electric motor. A packed section for the elimination of drift (or droplets of saline water) was installed above the coolant spray system in the air flow channel. The air flow rate was variable between 4 and 10 ft/sec by providing an adjustable gap in the cooling tower flow channel between the fan and the drift elimination packing.

Coolant cascading through the tower was collected in a first basin, with gravity flow into a second basin from a low, submerged outlet. Coolant was continuously recycled from an outlet adjustable to an intermediate level of the first basin and through a heat exchanger simulating the condenser of a large power plant, raising its temperature by 5 to 30°F before spraying the heated coolant into the top of the cooling tower. This heat exchanger comprised eight lengths of 90-10 copper-nickel condenser tubes of 20-ft length in a four-pass arrangement; this provided two parallel condenser tube channels of 80-ft length. Heat input was by rapid recirculation of water through the shell-side of this exchanger and four electric hot water heaters, each of 10-kW capacity.

Salinity control was provided by using a conductivity sensor actuating a feed pump to inject softened waste water into the top of the first basin when the conductivity of the recirculated coolant reached a preset maximal limit and to turn this pump off at a selected lower conductivity. Simultaneous overflow of concentrated coolant was thus provided for from the first basin into the second basin. The concentration of this cooling tower blowdown could thus be controlled at any desired TDS content, and the concentrate was automatically collected in the second basin for later use. A low-water shut-down probe installed in the first basin and a high-water shut-down probe in the second basin provided for the automatic tripping of all power to the facility, thus allowing continuous, unattended operation.

Thermocouples installed at 12 locations in the cooling tower loop were coupled with a stripchart recorder to provide a continuous record of all significant events in this loop. Liquid flowrates and temperatures were accurately measured to provide periodic heat-transfer performance data. The heat-transfer performance of the condenser was maintained during the first 6 weeks of operation of this pilot plant. No fouling of a biological or chemical nature was observed with the softened waste water coolant recycled at a 15-fold concentration level having a TDS of 54,000 ppm. Biological growth was controlled by an intermittent dosage of the coolant with sodium hypochlorite added to the contents of the first basin. Dechlorination was effected, as needed, by the addition of sodium sulfite. The heat exchanger performance was about 410 (±5) Btu/hr/ft^2/°F.

B. Brine Concentration by Vertical Tube Foam Evaporation (VTFE)

The objective of this operation was to further concentrate the cooling tower blowdown to 70,000 ppm of TDS, about 20-fold the original softened wastewater, sufficient to serve as the sole regenerant for the ion-exchange resin used in the softening pretreatment. A number of papers have been published on the development of VTFE[3,8,18] and several patents[19] have been issued on this method of increasing the rate of evaporation of liquids by induced foamy vapor-liquid flow (interface enhancement). A rotatably mounted upflow-downflow VTFE of 5000 gpd capacity was constructed for this facility, visible left of center in Figure 4. This procedure utilized 50 double-fluted aluminum-brass distillation tubes (Yorkshire Imperial Metals Ltd., England) of 6-ft heated length, 1.5-in. diameter, removably installed with "O"-ring seals. Stainless steel (type 316) was used for the main evaporator components and pumps. Instrumentation provided accurate temperature measurements on the feed (Tf), heating steam (Ts), and the brine (Tb) rejected from the distillation tubes. Typical operation of evaporators of this type, both by conventional and the VTFE method, in upflow and downflow modes, has been published elsewhere.[3,8,18-20]

The heat of vaporization to this unit was supplied from a 100 hp boiler, visible adjacent to the cooling tower in Figure 4. The steam supply was controlled by a reducing valve (90 psi to 45 psi), followed by a pneumatically operated steam control valve and a desuperheater to permit evaporation-concentration by VTFE at any temperature within the range of 100 to 250°F. The vapor produced was condensed in a large horizontal condenser and cooled with water of controlled temperature to maintain a steady temperature difference ($\delta T = Ts - Tb$) for evaporation. A large evacuable stainless steel tank was included in the VTFE loop to provide capability for the continuous concentration of 100-gal batches of cooling tower blowdown to acceptably low volumes. This blowdown was of sufficient concentration to serve as ion-exchange resin regenerant after first removing the interface-enhancement additive (foaming agent) and particulates from it by foam fractionation-flotation.[4]

VTFE heat transfer performance (U) was readily maintained at 2100 Btu/hr/ft²/°F, with a δT of 8.5°F, at an evaporation temperature of 220°F with the evaporator operated in the downflow mode, with 10 ppm of Neodol® 25-3A (Shell Chemical Company) added. The corresponding conventional performance (U), without a foaming agent additive, was 1400 Btu/hr/ft²/°F. These overall heat transfer coefficients were determined in the conventional manner, using the formulation $Q = UA\delta T$, where Q is the heat flux in Btu/hr, A is the tube outside surface area, and $\delta T = Ts - Tb$.

IV. CONCLUSIONS AND PERSPECTIVE

The results obtained with the pilot plant are promising. An adequate, sustained cyclic capacity of the ion-exchange resin for softening irrigation drainage water was observed. Use of the softened product in a cooling tower simulating actual process conditions in a large power plant cooling loop, at a 15-fold concentration factor, presented no apparent fouling or loss in heat-transfer performance during 6 weeks of continuous operation with intermittent chlorination. Further concentration of the cooling tower blowdown by VTFE to a final concentration factor of 20 was readily accomplished; a 50% improvement in the heat-transfer performance was obtained by the addition of 10 ppm of a selected surfactant. After removal of the surfactant by foam fractionation, the VTFE concentrate was sufficient as the sole regenerant for the ion-exchange resin.

Regeneration of the resin in the upflow, fluidized-bed mode utilizes formation of calcium sulfate crystals near the regeneration sites to drive this reaction towards completion while it removes the precipitate by hydraulic classification. Fluidized-bed regeneration requires a high flowrate, and the volume of fresh regenerant is limited by the volume of softened water produced per cycle. Under these conditions, the problem of providing sufficient contact time for regeneration has been solved by recycling a sufficient volume of stored regeneration effluent after separation of solid calcium sulfate by stirring and settling.

From these tests, use of irrigation-drainage water for power plant cooling appears technically feasible. Further tests in the field, extending over a 12-month seasonal cycle, are in progress and have confirmed these early data and observations. The results are intended to serve as the basis for an economic feasibility analysis of this promising water-resource-reclamation process.

With minor modifications, the ion-exchange softening method developed as pretreatment for evaporative processes should also be applicable as pretreatment for membrane desalination processes. High-sulfate, high-calcium waters, for which fluidized-bed regeneration is suited, occur over large areas of the U.S. and other countries, so that the method may find extended application.

ACKNOWLEDGMENTS

The process described was initiated and developed at the University of California Sea Water Conversion Laboratory, Richmond, as part of a 3-year project jointly sponsored by the California State Department of Water Resources (DWR), the Electric Power Research Institute, the Pacific Gas and Electric Company, the Los Angeles Department of Water and Power, and the Southern California Edison Company, and was coordinated by D. B. Brice of the DWR.

In addition to the financial support, the authors gratefully acknowledge many helpful suggestions by the sponsors' representatives, counsel by Professor Theodore Vermeulen of the University's Chemical Engineering Department, imaginative laboratory work by T. J. Jarvis, dedicated and resourceful technical support by J. C. Hensley and C. L. Freel, and efficient secretarial help by Rosemary Kirby and Amy Adelson.

The opinions expressed are those of the authors and not necessarily those of the sponsors.

REFERENCES

1. **Sephton, H. H. and Klein, G.,** A method of Using Irrigation Drainage Water for Power Plant Cooling, Proc. 1st Desalination Congr. American Continent, Mexico City, October 1976.
2. **Makar, K. M., Vermeulen, T., and Klein, G.,** Design and cost for ion-exchange softening for a seawater evaporation plant, in *Ion Exchange in the Process Industries,* Society of Chemical Industry, London, 1970.
3. **Sephton, H. H.,** Interface Enhancement for Vertical Tube Evaporation of Seawater, Proc. 4th Int. Symp. Fresh Water from the Sea, 471, Heidelberg, Germany, September 1973.
4. **Valdes-Krieg, E., King, C. and Sephton, H. H.,** Removal of surfactants and particulate matter from seawater desalination blowdown brines by foam fractionation, *Desalination,* 16(1), 39, February, 1975.
5. **Klein, G., Jarvis, T. J., and Vermeulen, T.,** A Method of Using Agricultural Waste Water for Power Plant Cooling. B. Reduction of Scaling Potential in High-Calcium, High-Sulfate Waters, Preliminary Final Report to the California State Department of Water Resources, November 1976.
6. **Haugseth, L. A. and Beitelshees, C. D.,** Evaluation of Ion-Exchange Pretreatment for Membrane Desalting Processes, Report RRC-ERC-74-26, Bureau of Reclamation, Department of the Interior, Washington, D. C.
7. **Taylor, I. and Haugseth, L. A.,** Yuma Desalting Plant Design, Proc. 1st Desalination Congr. American Continent, Mexico City, October 1976.
8. **Sephton, H. H.,** Recycle of Power Plant Cooling Tower Blowdown with Interface Enhancement, Proc. 2nd Natl. Conf. Complete Water Reuse, Chicago, May 1975.
9. **Carlson, R. S.,** Ion-Exchange Removal of Scale-Forming Components in Sea Water, M.S. thesis in Chemical Engineering, University of California, Berkley, 1965.
10. **Cherney, S.,** Scale Prevention Studies and Feasibility Analysis of Sea Water Softening by Fixed-Bed Ion Exchange, M.S. thesis in Chemical Engineering, University of California, Berkeley, 1966.
11. **Klein, G., Villena-Blanco, M., and Vermeulen, T.,** Ion-Exchange Equilibrium Data in the Design of a Cyclic Sea Water Softening Process, *Industrial Engineering Chemical Design and Development,* 3, 1964, 280.
12. **Klein, G., Cherney, S., Ruddick, E. L., and Vermeulen, T.,** Calcium removal from sea water by fixed-bed ion exchange, *Desalination,* 4, 158, 1968.
13. **Klein, G., Makar, M., Tleimat, B. W., and Vermeulen, T.,** Design and Cost of Ion-Exchange Softening for 50 mgd Sea Water Evaporation Plant, University of California Sea Water Conversion Laboratory Report 68-2, Part B, Richmond, 1968.
14. W. L. Badger Associates, Critical Review of Literature on Formation and Prevention of Scale, Office of Saline Water Research Development, Progress Report No. 25, U.S. Department of the Interior, Washington, D. C., 1959.
15. **Jarvis, T. J.,** Ion-Exchange Pretreatment of Brackish Water Using Fluidized-Bed Regeneration with Calcium Sulfate Precipitation, M.S. thesis, University of California, Berkeley, 1976.
16. **George, D. R., Riley, J. M., and Crocker, L.,** Preliminary Process Development Studies for Desulfating Great Salt Lake Brines and Sea Water, Bureau of Mines Report of Investigations 6928, U.S. Department of the Interior, Washington, D.C., 1967.
17. **Popper, K., Bouthilet, R. J., and Slamecka, V.,** Ion-exchange removal of sodium chloride with calcium hydroxide as recoverable regenerant, *Science,* 141, 1038, 1963.

18. **Sephton, H. H.,** Desalination by Upflow Vertical Tube Evaporation with Interface Enhancement, Proc. Intl. Desalting Environmental Assoc. Conf., Ponce, Puerto Rico, April 1975.
19. **Sephton, H. H.,** Interface Enhancement Applied to Evaporation of Liquids, U.S. Patent No. 3,846,254, November 1974, and foreign patent counterparts.
20. **Sephton, H. H.,** New Developments in Vertical Tube Evaporation of Seawater, Proc. 5th Int. Symp. Fresh Water from the Sea, Vol. 2, 1976, 279.

Chapter 7

MUNICIPAL WASTE EFFLUENT TREATMENT

I. R. Higgins and R. C. Chopra

TABLE OF CONTENTS

I. INTRODUCTION

The treatment of water by ion exchange is a well-known unit operation and has been used in industry for boiler feed and process water treatment for almost half a century. The major operating cost is for regenerant chemicals used in greater than the stoichiometric amount of the contaminants to be removed. Ion exchange is economical compared to precipitation, solvent extraction, or other treatment processes, provided the required excess over the stoichiometric amount of chemicals for regeneration of the ion exchange resin can be kept to a minimum and the regenerant waste is disposable. The conventional fixed-bed ion-exchange equipment in wide use for many years traditionally uses large quantities of regenerants, many times in excess of that required by stoichiometry. The regenerants must also form soluble salts to avoid precipitation inside the matrix of the resin bead or to avoid plugging of the bed. This restricts the use of the lowest cost acids and alkalis, for example, sulfuric acid and lime.

Some parts of the U.S., Southern California, for example, long faced with a deficit of good quality fresh water, have available a potential source in its once-used wastewater. An efficient, economical, and reliable demineralization process which can be adapted to the special problems encountered in wastewater treatment would eliminate one of the obstacles in the utilization of several million acre feet of wastewater now being discharged annually to the oceans.

The mineral quality of the wastewater varies considerably, from about 600 mg/ℓ total dissolved solids (TDS) to ca. 3000 mg/ℓ. Water quality surveys indicate that ca. 50% of the wastewaters have a concentration of less than 1000 mg/ℓ.

The ion-exchange process, which has been successfully employed for many years for municipal and industrial purposes, has only recently been considered for demineralization of wastewater. It is believed that blending of the demineralized wastewater with nondemineralized wastewater to produce a preselected quality of water will prove economical. Using this approach, cost estimates have been prepared which propose the removal of ca. 300 mg/ℓ TDS, which is generally considered the increment added during one domestic use cycle of a water supply.[1]

Partial demineralization of secondary effluents from a sewage treatment plant following filtration, etc., is considered a possible method to produce potable water,[2] if used in conjunction with other processes to remove, for example, organics. It could also be used by itself to produce high quality industrial water. The use of ion exchange for selective removal of certain nutrients like ammonium nitrate and/or phosphate from effluents is also being studied.[3]

II. DEMINERALIZATION OF SECONDARY EFFLUENTS

A. Fixed-Bed Column Operation

In the usual fixed-bed system, columns of cation and anion exchange resins are placed in series as shown in Figure 1. The resins do not move during the service, regeneration and rinse operations. Cation exchange is accomplished using a strong acid resin in the hydrogen form while weak base resins in the free base form are employed for anion exchange. The effluent quality characteristics of a fixed-bed ion exchange pilot plant are shown in Table 1.

The anion resin acts as a weak base resin when regenerated with ammonium hydroxide and does not remove weak acids. For this reason, no reduction in silica takes place; it would also be expected that bicarbonate would not be removed. However, a considerable reduction in alkalinity has been observed.

The product water pH of 5.8 is lower than desired, but it is anticipated that the effluent from the ion exchange process will be blended with nondemineralized wastewater prior to reuse. The buffer capacity of the ion exchange product water is quite low when compared to the waters with which it will be blended. Tests confirm that the pH of the ion exchange product water will be increased to the pH of the water with which it is blended.

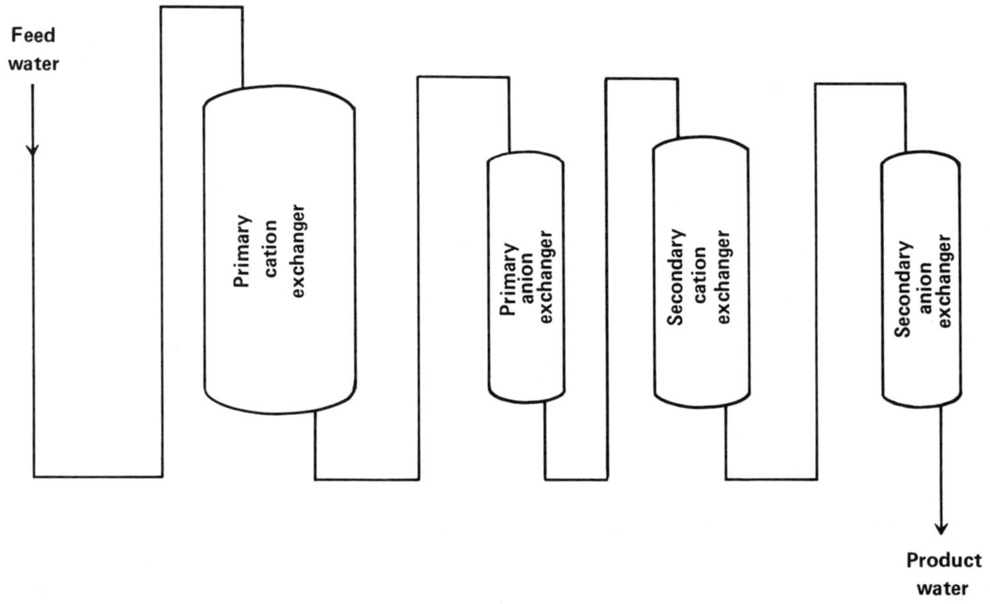

FIGURE 1. Demineralization of secondary effluent. Feed water = effluent from activated carbon column.

TABLE 1

Average Water Quality Characteristics of a Fixed-bed Ion Exchange Pilot Plant Under Typical Operating Conditions

Constituent[a]	Carbon column effluent (feed influent)	Primary cation effluent	Primary anion effluent	Secondary cation effluent	Secondary anion effluent (product)
Calcium	53	2.0	1.7	1.1	0.60
Magnesium	17	0.59	0.56	0.38	0.00
Sodium	126	61	59	16	15
Potassium	14	7.3	7.1	1.9	1.9
Ammonia, as N	20	9.6	9.2	4.0	1.3
Sulfate	72	72	3.6	3.6	1.3
Nitrate, as N	2.9	2.8	1.6	1.5	0.35
Chloride	135	132	84	83	14
Orthophosphate, as PO_4	27	27	15	14	0.25
Total alkalinity, as $CaCO_3$	218	—	51	—	39
H^+ as $CaCO_3$	—	110	—	110	—
pH	7.4	2.7	5.7	2.8	5.8
COD	10	9.7	6.8	5.5	3.7
Silica, as SiO_2	23	—	—	—	23
TDS	610	298	198	104	72
Conductivity, μmhos/cm	1040	1039	390	1040	100

[a] All constituents in mg/ℓ except pH and conductivity.

From Parkhurst, J. D., Chen., C., Carry, C. W., and Masse, A. N., *Advances in Water Pollution Research,* Vol. 1, Jenkins, S. H., Ed., Pergamon Press, New York, 1971. With permission.

Some reduction in chemical oxygen demand (COD) occurs during passage through the resin. A chemical balance around the plant indicates that the majority of COD removed can be accounted for in the reject regenerant and backwash streams.

B. Continuous Countercurrent Ion Exchange System

The Chem-Seps Continuous Countercurrent Ion Exchange (CCIX) system* overcomes many of the deficiencies of fixed-bed systems because of the inherent advantages of continuous and countercurrent flows and exploits the use of the lowest cost regenerant chemicals.[4] Some of the advantages of CCIX are

1. Maximum utilization of regenerant chemicals by continuous countercurrent flow.
2. Slurries and precipitates may be handled in a dynamic flow system where there is high production rate per unit equipment size and no stagnant flow pockets. This allows the use of low cost H_2SO_4 and lime as regenerants.
3. Weakly ionized H_2SO_4 is made highly efficient by using a mixture of weak and strong acid type resins.

The series of available cost-saving features of CCIX make ion exchange the most economical method for the treatment of secondary effluents. Previous work indicates that in this system, especially with a high bicarbonate water, sulfuric acid and lime are not only successfully handled but are required in close to stoichiometric amounts.

* Covered by U.S. Patent 3,580,842 and assigned to the Chemical Separations Corporation.

Continuous countercurrent ion exchange has been successfully used for several pollution control applications.[4] However, ion exchange also increases the total amount of solids in the regenerant waste streams. But for many applications and with the proper choice of regenerant chemicals, the concentrated waste stream can be recovered as a by-product.

1. System Operation

The CCIX system is a single loop in which various sections are separated by valves to isolate the sections performing different functions simultaneously. To explain the operation of a typical unit, a simple configuration used in water softening or demineralization may be used (see Figure 2). The resin is moved in small increments by the action of a hydraulic ram. All solution flows are countercurrent to the flow of resin. The contaminants are removed in the process cycle, and the resin is moved in the pulse cycle. A controller in the effluent stream from the treatment section senses the upper limit permissible in the effluent, for example, conductivity. In the process cycle, the influent enters through a set of collectors. Simultaneously, the regenerant enters through valve RE and leaves through valve WA. The rinse enters through valve RI, and its duration is controlled by an interface controller. In the process cycle, valves B, C, and D are closed, isolating all the sections. The duration of the process cycle is 4 to 10 min, depending on the contaminants to be removed, and is controlled either by a timer, if the contaminants are fixed, or a controller in the effluent stream, if the contaminants fluctuate at short intervals.

Valve A is opened, and resin is dropped to the pulse section simultaneously with the removal of contaminants in the treatment section, the regeneration of loaded resin in the strip section, and the rinsing of regenerated resin in the rinse section. As the resin drops through the valve A, it is backwashed to remove the suspended solids or resin fines. Removal of these fine particles is important to avoid the unnecessary pressure drop in the bed and plugging of the distributors.

At the end of the process cycle, all flows going into and out of the unit are automatically shut off. Valve A closes and valves B, C, and D open in the proper sequence. A slug of water enters through valve PU, a predetermined incremental volume of resin moves into the regeneration section, an equal volume of regenerated and rinsed resin enters the bottom of the treatment section, and another equal volume of the loaded resin leaves the top of the treatment section and enters the resin reservoir section. The resin moves in a plug flow that is laminar; no fluidization occurs during the pulse cycle. The total time required for proper sequencing of valves and the movement of resin is 20 to 30 sec. At the end of this period, valves B, C, and D close, and valve A opens. The solution valves open, and normal steady operation begins. All valves are interlocked and properly sequenced. The resin movement is precisely controlled by sonic devices.

2. Special Features

a. Downflow

The raw feed in the treatment section moves downward, and the resin moves upward in the treatment section. The "downflow" of feed overcomes the handicap encountered in "upflow" systems. The upflow of feed requires a certain minimum velocity during the process cycle to lift the bed and keep it in suspension. If a certain velocity is not maintained, resin begins to churn in the loading section. In the downflow system, resin drops in the treatment section by gravity and the feed flow in the downward direction further keeps the resin settled. The maximum velocity in downflow operations is 60 gpm/ft^2. There are no limitations on the minimum velocity as compared to a minimum velocity of 20 to 30 gpm/ft^2 in upflow systems.

Another advantage of downflow is the use of raw water during pulse and backwash.

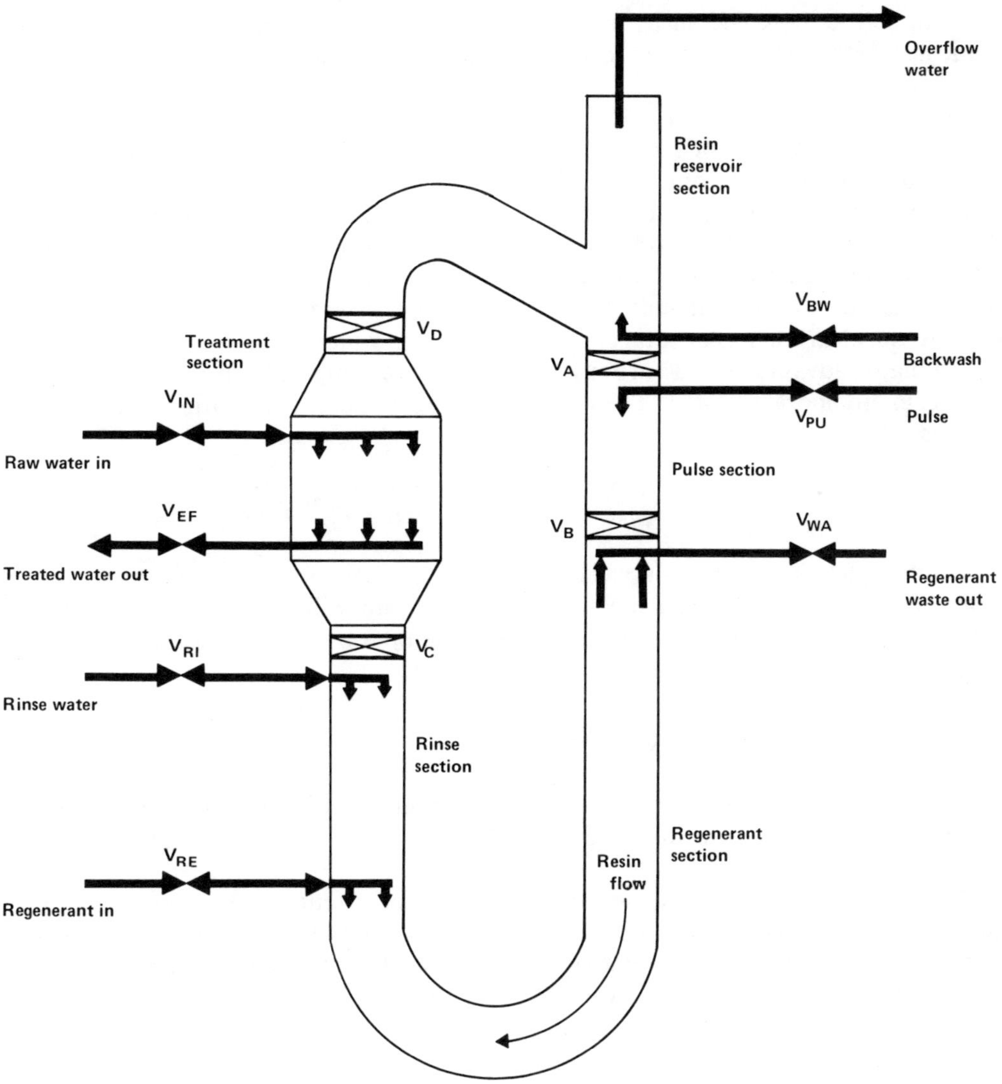

Valve positions during cycles

Run cycle		Pulse cycle	
Valves open	Valves closed	Valves open	Valves closed
A,WA,IN,EF RI, RE	B,C,D,PU	B,C,D,PU	A,WA,IN,EF,RI,RE

FIGURE 2. Schematic diagram of the continuous countercurrent ion exchange (CCIX) contactor.

The water which is used for pulse goes out in the waste. The resin in the pulse section is fully loaded, whereas in an upflow system the resin in the pulse section is fully regenerated and rinsed, with treated water. The only treated water requirement in the CCIX system is the rinse water which in volume is equal to the volume of resin moved. Another significant advantage of downflow operations is that the liquids with higher specific gravity are always at the bottom part of the loop, thus making it easier to control the interfaces and obtain concentrated wastes.

The performance of a downflow system using low density resins is considered superior to upflow performance with similar resins. A weak base resin like Dowex® MWA-

1 or Amberlite® IRA-93 in hydroxide form when transferred to the pulse section is in the heavier sulfate or chloride form, and settling presents no problems.

b. Ion Exchangers

The demineralization of secondary effluents requires three CCIX contactors, containing both weak and strong acid cation exchangers and weak base anion exchangers. Raw water is fed from a pump station to the weak acid cation exchanger loop; this is the first step in the demineralization process. The weak acid unit is designed to remove total hardness (Ca^{++} and Mg^{++}) equivalent to the bicarbonate ions. Normally, weak acid resins cannot split salts but the carboxylic type of weak acid resins will split salts equal to the bicarbonates or total alkalinity present in water.

The major advantage of a carboxylic resin is its very high affinity for the hydrogen ions from any inorganic acid. The ion exchange reaction in the treatment section is as follows:

$$2R^*{-}H + \binom{Ca}{Mg}(HCO_3)_2 \rightleftharpoons \begin{matrix}R^*\\R^*\end{matrix}\!\!>\!\binom{Ca}{Mg} + 2H_2CO_3$$

where the asterisk designates a weak-type resin, i.e., either a weak acid or weak-base resin. The spent regenerant from the strong acid unit will be used to regenerate the weak acid resin, reducing the sulfuric acid consumption. The ion exchange regeneration reaction for the weak acid resin is shown below:

$$\begin{matrix}R\\R\end{matrix}\!\!>\!\binom{Ca}{Mg} + H_2 2SO_4 \rightleftharpoons 2R{-}H + \binom{Ca}{Mg} SO_4$$

The strong acid unit will reduce the residual cations to 50 ppm. The cations removed in this unit, in addition to the residual Ca^{++} and Mg^{++} ions, are essentially the sodium ions.

A decarbonator is required between the strong acid unit and the weak base anion unit. The purpose of the decarbonator is to remove the CO_2 from the water. The weak base anion exchange unit will remove all the free acids from the decarbonator effluent. The ion exchange reaction in this treatment section is as follows:

$$R^*{-}H{-}OH + HCl \rightleftharpoons R^*{-}HCl + HOH$$

$$2R^*{-}H{-}OH + H_2SO_4 \rightleftharpoons \begin{matrix}R^*{-}H\\R^*{-}H\end{matrix}\!\!>\!SO_4 + 2HOH$$

The pH of the water coming from the weak base unit will be in the range of 6 to 8. The loaded resin is regenerated with the recovered NH_4OH. The reaction is as follows:

$$R^*{-}HCl + NH_4OH \rightleftharpoons NH_4Cl + R^*{-}H{-}OH$$

$$\begin{matrix}R^*{-}H\\R^*{-}H\end{matrix}\!\!>\!SO_4 + 2NH_4OH \rightleftharpoons (NH_4)_2SO_4 + 2R^*H{-}OH$$

3. Costs

The economic evaluation to be presented is based on the objective of maximum TDS removal at the lowest cost per ton of salts removed. To achieve this objective, the proposed system is designed to remove 950 ppm TDS from the raw water containing 1000 ppm TDS.

The design installed capacity of the system considered will be 81.4 million gallons per day (mgd). There will be two spare trains available at all times. The treated water supplied will be 75 mgd based on the requirements of a reasonably sized city. Before presenting the economic evaluation, a number of cost factors which have a significant influence on on the interpretation of the economic data have been listed below:

1. Raw water charge at $120/mgd (includes pretreatment costs)
2. Suspended solids disposal at $17/ton
3. Land cost at $30,000/acre
4. Direct and indirect labor and overhead: The direct labor includes four operators and one supervisor per shift. Each operator will monitor two modules. The indirect labor includes three maintenance persons and two technicians on the day shift only. The overhead included is 50% of the direct labor costs. A central control room is included to keep the labor to a minimum.
5. Resin replacement: An average life of 5 years has been used for the resin. The cost used is the average cost for all three types of resins. The costs used do not take into consideration the large-quantity discount which should be available for a yearly consumption of 11,430 $ft.^3$

A summary of the capital and operating costs is given in Tables 2 and 3. The operating cost is 42¢ per 1000 gallons or in terms of salt removal, the cost is $93 per ton of salt removed.

III. SELECTIVE NITRATE REMOVAL FROM POTABLE WATERS

In recent years, nitrate contamination has become increasingly critical in many areas of the U.S. as a result of cesspool effluents and fertilizer run-off into both groundwater aquifers and surface waters. Specifically, nitrate levels have been increasing at an alarming rate in the eastern states. The ingestion of water high in nitrates can cause methemoglobinemia, a condition in infants under one year of age. For this reason, the drinking water standards have set a limitation of nitrate nitrogen (NO_3-N) at 10 ppm (mg/ℓ).

Continuous ion exchange is recommended in this application in order to achieve high production rates in a full-scale water treating plant and to gain higher efficiencies in the operating costs, particularly in regenerant salt consumption.[5] The basic design criteria for sizing a continuous ion exchange plant for nitrate removal are

1. A flowrate of 50 gpm/ft^2 of cross-section in the nitrate removal section of the ion exchanger
2. A NO_3-N resin utilization of 0.4 eq/ℓ
3. Salt consumption for 10 ppm NO_3-N removal (from 15 ppm to 5 ppm) of 1.0 lb/1000 gal, for 20 ppm NO_3-N removal (from 25 ppm to 5 ppm) of 2.5 lb/1000 gal, and 5.0 lb/1000 gal for 50 ppm NO_3-N removal (from 50 ppm to less than 10 ppm)

Figure 3 is a schematic flowsheet of a 2 mgd, 20 ppm NO_3-N plant using a strong base anion exchange resin in the chloride form. Table 4 contains the basic operating costs and approximate capital costs for a 2 mgd plant with three levels of nitrate removal. The operating costs include amortization but not interest costs. The plant is assumed to operate at a 50% load factor (average output is 1 mgd based on the quanity of water blended and the yearly demand. Figure 4 shows costs as a function of nitrate removal.

TABLE 2

Capital Costs of a 75 mgd Demineralization Plant

TDS	
Influent ppm	1000
Effluent ppm	50
Plant size (mgd installed)	81.2
Load factor (%)	100
Product water recovery (%)	95
Treated water (mgd)	75

Expenditures	Unit cost	Total cost
A. Materials		
1. Complete desalting module	—	$8,511,000
Individual units (75)		
Ion exchange trains (25)		
2. Instrumentation	—	625,000
3. Feed pretreatment system	None	
4. Auxiliary equipment:	$858,000	
a. Feed and transfer pumps	375,000	
b. Decarbonators and clearwell	200,000	
c. Tanks H_2SO_4 storage and transfer system	30,000	
d. Acid dilution and reuse systems	59,400	
e. Resin recovery tanks	93,600	
f. Soda ash system	100,000	
5. Waste Neutralization		1,000,000
Equipment Cost		$10,094,000
B. Installation		$1,305,000
1. Equipment	500,000	
2. Piping	555,000	
3. Electrical	250,000	
C. Civil work		670,000
1. Land (5 acres)	150,000	
2. Site work including roads	150,000	
3. Building, 10,000 ft^2: (Office 2000 ft^2; laboratory 600 ft^2; central control room 15000 ft^2; maintenance and warehouse 5900 ft^2)	225,000	
4. Foundation pads for equipment, pumps, tanks and draining facilities	145,000	
D. Start-up		200,000
Subtotal		$12,269,000
E. Contingency 5%		672,000
Total Capital Cost		$12,941,000

Note: All costs based on 1973 values.

TABLE 3

Operating Costs of 75 mgd Demineralization Plant

Expenditures	Units	Unit cost	Dollars per day
Fixed expenses			
Amortization (6% for 20 yr)	—	—	$3,048
Indirect labor and overhead	—	—	222
Material and supplies	3% of Equipment cost per year	—	829
	Subtotal		$4,099
Variable expenses			
Direct labor and overhead	5 Men/shift	—	1,000
Chemicals			
Sulfuric Acid	Ton	$ 15.30	4,071
Soda Ash	Ton	$ 35.40	8,602
Power	KWH	1¢	750
Feed water	mg	$120	9,480
By-product disposal	Solids-Ton	$ 17	
	Liquid-mg	$ 33	2,848
Replacement Ion exchange resins	ft^3	$ 26.25 (average)	822
	Subtotal		$27,573
	Total Operating Cost		$31,673[a]

[a] 42.22¢/1000 gal, or $92.61/ton of total dissolved solids removed.

IV. CONCLUSIONS

Ion exchange has been applied for many years to many municipal water supplies for softening and iron removal. Fluoride removal by means of apatites, which can exchange OH for F, came into use in the 1950s in some areas with waters containing high fluoride contents. In recent years, the removal of nitrate by means of strong base anion exchangers in the chloride form is being practiced in a limited number of communities. Desalination of brackish waters by ion exchange membranes (electrodialysis) is practiced in some southern states.

In municipal waste effluent treatment, ammonia removal by an inorganic cation exchanger is practiced in a few communities. However, cost is a major consideration by municipalities when it comes to applying new technologies. Necessity and legislation are forcing communities to consider new technical developments. Thus communities with traces of toxic heavy metals and/or traces of carcinogenic organics will lead to demands by citizens for their removal. Thus, specific exchangers, polymeric adsorbents, extracting resins, etc., will come under consideration. Thus, the future for use of polymers with active sites holds many possibilities for municipal application as there will be demands for making drinking water safe and pleasant to drink and for making waste effluents safer to be discharged or to be reused.

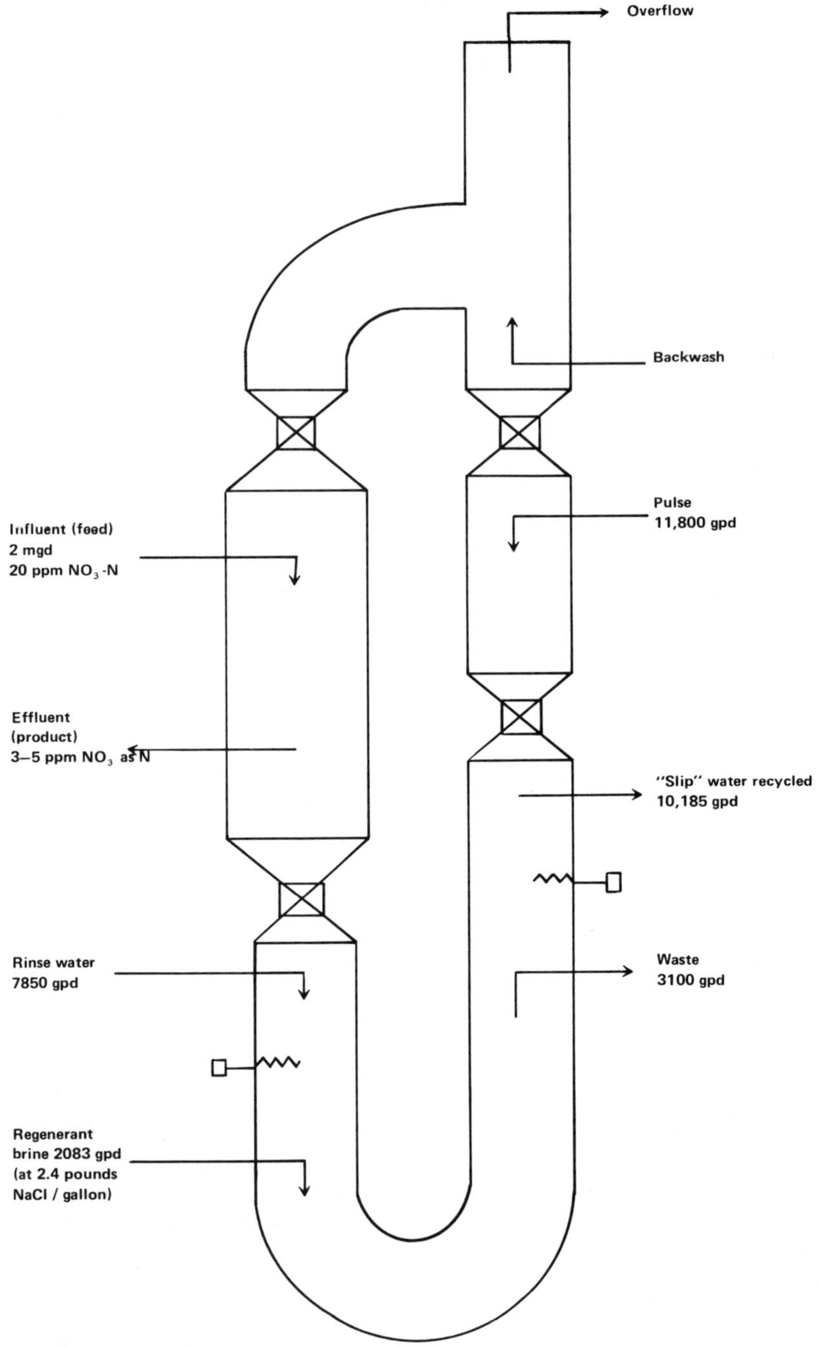

FIGURE 3. Schematic flow sheet of a 2-mgd unit with 20 ppm NO_3-N removal. For 50% load factor, waste volume will be 1550 gpd; brine consumption, 1042 gpd; and rinse, 3925 gpd.

TABLE 4

Operating and Amortization Costs in Dollars/1000 gal of a 2 mgd CCIX NO_3-N Removal Plant Operating at a Load Factor of 50% (Average Daily Output, 1 million gallons)

	10 ppm NO_3-N Removal	20 ppm NO_3-N Removal	50 ppm NO_3-N Removal
NaCl regenerant salt costs (ca. $14/ton)	0.0070	0.0175	0.0350
Resin replacement, minimum life of 5 years	0.0106	0.0106	0.0156
Other operating costs—labor, maintenance, utilities	0.03666	0.03666	0.03666
Total operating costs	0.0542	0.0647	0.0872
Amortization — 20-year straightline (no interest included, total capital cost $285,000 to $320,000)	0.0390	0.039	0.0438
Total cost (operating and amortization)	0.0942	0.105	0.1324

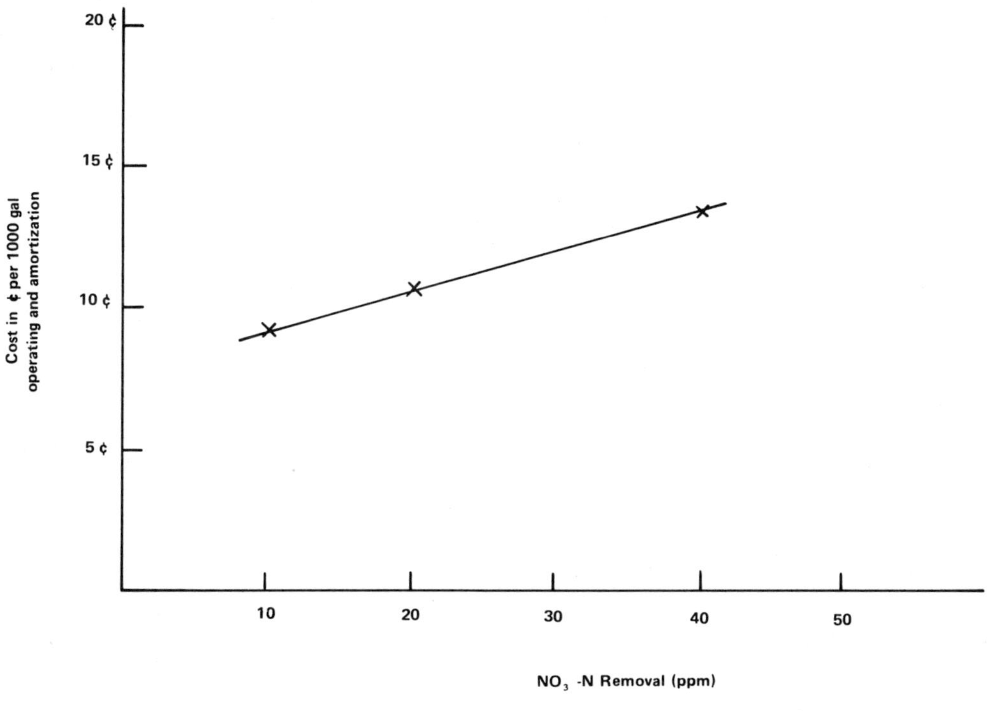

FIGURE 4. Operating and amortization costs of a 2-mgd CCIX NO_3-N removal plant.

REFERENCES

1. **Higgins, I. R.**, Ion exchange: its present and future use, *Environ. Sci. Technol.*, 7(13), 1110, December 1973.
2. **Parkhurst, J. D., Chen, C., Carry, C. W., and Masse, A. N.**, Demineralization of wastewater by ion exchange, in *Advances in Water Pollution Research*, Vol. 1, Jenkins, S. H., Ed., Pergamon Press, New York, 1971.
3. **Holzmacher, R. G.**, Nitrate removal from a ground water supply, *Water Sewage Works*, 118(7), 210, 1971.
4. **Higgins, I. R., and Chopra, R. C.**, Chem-Seps Continuous Ion-exchange Contactor and its applications to de-mineralization processes, in *Ion Exchange in the Process Industries*, Society of Chemical Industry, London, 1970, 121.
5. **Gregg, J. C. and Bowne, S. B., Jr.**, Nitrate removal from municipal well water, *Proc. 34th Int. Water Conf.*, Engineers' Society of Western Pennsylvania, Pittsburgh, 1973, 179.
6. **Bingham, E. C. and Chopra, R. C.**, A unique closed cycle water system for ammonium nitrate producers, Proc. 32nd Int. Water Conf., Engineers' Society of Western Pennsylvania, Pittsburgh, 1971, 143.

Chapter 8

REGENERANT WASTES

C. Calmon

TABLE OF CONTENTS

I. INTRODUCTION

While ion exchange is used as a tool in reducing pollution, it can produce polluting effluents if spent regenerant cannot be utilized. This is especially true in water softening and in water demineralization. In the U.S., there are seven million water softeners, which use an average of about 150 lb of salt per year for regeneration. This gives an annual use of over a half million tons, of which half is not utilized and which is discarded into sewage lines, streams, or is injected into the ground. Legislation is already limiting the disposal of regenerant streams in those areas where water supplies are scarce and/or brackish.

Thought is being given to solving the problem in various ways. Some solutions to the problem are still in their infancy, while others are based on the full utilization of the salt to reduce the quantity presently used. The latter can also represent a saving in operating costs. This aspect of the problem has been emphasized more in Europe than in the U.S. The various reductions of pollution due to regenerant wastes are described in the following sections.

II. METHODS FOR REDUCING REGENERANT WASTES

A. Elimination of Chemicals as a Regenerant

1. Water as an Eluting Agent

Water as an eluting agent can be utilized in many processes. Since this is a new development with many possibilities, a separate chapter (Volume II, Chapter 9) covers this subject.

2. Membrane Processes

Electrodialysis — An electric current and ion-exchange membrane are used for the removal of the polluting ions.[1] Since it is more economical to reduce the ion concentration by electrodialysis to about 500 ppm, ion exchange becomes practical in reducing the ionic content to lower values.[2] Dow has applied an electrodialysis process for reducing the volume of cooling tower discharge as well as boiler blowdown.[3] The concentrated stream can be readily reduced to dryness. If the salts are mainly sodium, then they could be used as a regenerant for softening the makeup water by ion exchange.

Reverse osmosis - The rejection stream will contain 90 to 95% of the ions.[4] The residual ions in the water can be reduced by ion exchange to any value desired. In this manner the amount of regenerant is limited for the removal of the residual ions.

Ultrafiltration - Particulates and high-molecular-weight entities can be removed, so there is no need for chemical pretreatment essential to prevent fouling of membranes or ion-exchange beds.

3. Electroprocesses

The Office of Saline Water, now the Office of Water Resources and Technology, has investigated several electroprocesses for desalination. As yet, although feasible, the processes have not been sufficiently economical to be applied. One process from the University of Arizona utilizes electrodes composed of polymeric materials with polar sites with high affinities for the salt form, but which are converted to the H^+ and OH^- forms when a direct current is applied. Another process involves a mixture of cation and anion exchangers placed between electrodes. The exchangers in the salt forms are converted to the H^+ and OH^- forms by an applied current, i.e., regeneration of the ion exchangers is accomplished by means of a current. This work was done at the Southern Research Institute, Montgomery, Alabama.[6]

B. Reduction of the Regenerant Dosage

Reuse of the least efficient portion of the regenerant — As the regenerant passes through a bed of ion exchangers, the efficiency is reduced because the equilibrium conditions are constantly changing in the bed. Thus, the last portion of the effluent contains a great deal of unutilized regenerant. This part can be saved and reused in the next cycle prior to the use of fresh regenerant. The recovered portion of the regenerant can be saved in a tank or through the use of multiple beds.[7]

Counter-current operations — Counter-current operation produces a better quality of effluent and can be operated in such a manner as to utilize the regenerant more efficiently thereby reducing the chemical usage and the quantity of spent regenerant.[8]

Use of weak exchanger-types — Because of the high affinity of carboxylic acid resins for the hydrogen ions, and weak-base anion exchangers for the hydroxide ion, almost stoichiometric regenerant usage can be obtained. These exchangers can be combined with the strong-type exchangers to reduce the quantity of regenerant. Carboxylic acid exchangers are utilized for the removal of the bicarbonate ion, while the weak-base anion exchangers are utilized for the removal of the strong acids, thus leaving a smaller number of ions to be removed by the strong-type exchangers for which the regenerant efficiency is lower.

Lime precipitation of calcium carbonate —When bicarbonate is present, it reduces the quantity of salts to be removed by the ion exchangers, as can be seen from the following equation:

$$Ca^{2+} + 2HCO_3^- + Ca(OH)_2 \rightarrow 2CaCO_3\downarrow + 2H_2O \qquad (1)$$

The precipitate can be readily disposed.

Dealkalization and softening with a carboxylic acid resin — Dealkalization and softening with a carboxylic acid resin in the H form reacts with bicarbonate to form H_2O and CO_2, the latter being removed by a degasifier. At the same time an equivalent amount of hardness or sodium goes on the ion exchanger. The regeneration of the exchanger with an acid occurs close to stoichiometry. Thus, dealkalization, softening, and partial deionization are obtained with a minimum of regenerant.[9]

Continuous counter-current operation — Although continuous counter-current operation is not used to a great extent in the U.S. due to many factors, it is in operation in many countries in Europe with excellent results, i.e., with high regenerant efficiencies and low ion exchanger inventory.[10]

C. Utilization of Waste Chemicals as Regenerants

Effluents from electrode chambers — Effluents from the electrode chambers in electrodialysis will contain either caustic or acid. These can be used to dealkalize and soften the influent water with a carboxylic acid resin in order to reduce fouling of the membranes. Thus there is a saving in the chemicals used as the regenerant.[11]

Reverse osmosis reject stream — The reject stream in some reverse osmosis systems may be utilized for regenerating the cation exchanger used for softening the incoming water to prevent membrane fouling.[12]

Seawater as a regenerant — Seawater may be used as a regenerant for cation exchangers if hardness leakage can be tolerated. If necessary, an additional unit regenerated with fresh regenerant may be used for scavenging.[13]

Sodium sulfate waste — Waste sodium sulfate may be used as a regenerant for softeners, provided it is initially used in dilute solution to prevent calcium sulfate precipitation. A calcium sulfate precipitation inhibitor may be used.

D. Treatment of Waste Effluents

Disposal of nontoxic salts is permissible in some streams provided there is enough dilution, no change in pH, and no precipitate formation. Regenerant effluents from a cation exchanger are usually acidic, while the effluents from an anion exchanger are alkaline. One procedure developed by Cochrane and described in Applebaum's text[14] is to remove the acid with a carboxylic acid resin in the salt form:

$$H^+ + \text{Na carboxylic resin} \rightarrow Na^+ + \text{H carboxylic resin} \qquad (2)$$

The effluents from the anion exchanger are passed through the acidic bed and the excess caustic is neutralized.[15]

$$\text{H carboxylic} + Na^+ \, OH \rightarrow \text{Na carboxylic} + H_2O \qquad (3)$$

E. Use of Specific Ion Exchangers

Volume II, Chapter 15, in this book deals with this subject.

F. Volatile Regenerants which are Recoverable

While attempts have been made to use volatile compounds as regenerants, e.g., CO_2 and NH_3, with a view towards recovery for reuse, thus far they have not found industrial applications, although a great deal of ingenuity went into the research work.

1. Use of CO_2 as a Regenerant

The carboxylic group has a high affinity for the hydrogen ion. When a carboxylic acid resin in a salt form is contacted with a solution containing CO_2, the exchanger hydrolyzes to the H form and the $NaHCO_3$ salt is formed. However, an equilibrium relation is maintained unless the $NaHCO_3$ is removed. The CO_2 concentration is increased with the gas under pressure.

$$RCOONa + H \cdot HCO_3 \rightleftharpoons RCOOH + NaHCO_3 \qquad (4)$$

The slight regeneration of carbonacous cation exchangers with CO_2 was observed very early without the realization that the COO^- groups had to be present in the exchanger. The degree of COO^- group formation on sulfonation of coal depended on the sulfonating agent.

Later, when exchangers with only carboxylic acid groups were prepared, the work was repeated, but there was limited application. Still later, the Desal process (cf. Volume I, Chapter 9) was developed in which CO_2 was used to convert a weak-base anion exchanger to the bicarbonate form, which can exchange the chloride and sulfates[16]

$$R \cdot N + H_2CO_3 \longrightarrow R \cdot \overset{H}{N}HCO_3$$

$$\downarrow \; Cl^- \; [\text{or } SO_4^=]$$

$$R \cdot \overset{H}{N}Cl \; [\text{or } (RN)_2SO_4] + NaHCO_3 \qquad (5)$$

The $NaHCO_3$ effluent is passed through a hydrogen carboxylic acid exchanger forming H_2CO_3 in solution.

$$RCO_2H + NaHCO_3 \longrightarrow RCO_2 + H_2CO_3$$

$$\downarrow$$

$$H_2O + CO_2 \uparrow \qquad (6)$$

The CO_2 can then be recycled for converting the base exchanger to the bicarbonate form.

2. Use of Ammonia as a Regenerant

Ammonia has been used for many years to regenerate weak-base anion exchangers. The waste effluent contained ammonium salts which were discarded. Since ammonia is a nutrient for organisms, its use has become limited. In the Desal process, ammonia is recommended as the regenerant for the anion exchanger with a view towards recovery for reuse.

At the University of California, Berkeley, a great deal of work has been done on brackish water with the emphasis on the recovery of regenerants.[17] However, the system is too complex and energy costs for volatilization of the gaseous regenerants have increased in recent years.

These examples indicate that efforts are being made to reduce regenerant wastes through recovery of the regenerant for reuse.

G. Regenerant Recovery as a Solid By-product

DePree and co-workers at Aerojet General developed a process which combines recovery and recycle of the regenerants and the waste salts as precipitates.[18] The process has been applied for deionization of brackish waters, industrial waste streams, and secondary sewage effluents. The chemicals used are organic acids and bases which are separated and recovered for reuse by distillation. The process, however, is expensive with a cost of about \$5/1000 gal of solution treated.

REFERENCES

1. **Schaffer, L. H., and Mintz, M. S.,** Electrodialysis, in *Principles of Desalination,* Spiegler, K. S., Ed., Academic Press, New York, 1966, Chap. 6.
2. **Heit, A. H., and Calmon, C.,** Electrodialytic recovery of sulfuric acid and iron, in *Membrane Processes in Industry and Biomedicine,* Bier, M., Ed., Plenum Press, New York, 1971, 145.
3. **Jordan, D. R., McElhenny, W. F., and Westbrook, G. T.,** Cooling tower effluent reduction by electrodialysis, *Proc. Power Conf.,* 38, 980, 1976.
4. **Lacey, R. E. and Loeb, S., Eds.,** *Industrial Processing with Membranes,* Interscience, New York, 1972.
5. **Eykamp, W.,** Ultrafiltration of Aqueous Dispersions, in 79th Natl. Am. Institute Chem. Eng. Meeting, Houston, Texas, March 1975.
6. **Davis, T. A. and Lacey, R. E.,** Electroregeneration of Ion-exchange Resins, Report to National Institutes of Health, Bethesda, AK-2-70-2108-F, 1972.
7. **Adams, R. C.,** Progressive mode continuous countercurrent ion exchange, Proc. 31st Int. Water Conf., Engineers Society of Western Pennsylvania, Pittsburgh, 1970. 1970.
8. **Abrams, I. M.,** Countercurrent fixed-bed Ion Exchange, 76th Natl. Meeting, Am. Institute Chem. Eng., Tulsa, OK, 1974.
9. **Applebaum, S. B.,** Application of carbonaceous zoolites to water softening, *J. Am. Water Works Assoc.,* 30, 947, 1938.
10. **Gold, H. and Todisco, A.,** Wastewater reuse by continuous ion exchange, in *Complete Water Reuse,* Cecil, L. K., Ed., AICRE, New York, 1973, 96.
11. **Halpern, B. M. and Oilie, J.,** 1.25 mgd electrodialysis plant in Israel *J. Am. Water Works Assoc.,* 64, 735, 1972.
12. **Kinsbury, W., and Schomaker, W. P.,** Proc. 34th Int. Water Conf., Pittsburgh, PA, 99, 1973.
13. **Richheimer, C. E.,** Sarasota's automatic sea water regeneration zeolite water softening plant, *J. Am. Water Works Assoc.,* 29, 1712, 1973.
14. **Applebaum, S. B.,** *Demineralization by Ion Exchange,* Academic Press, New York, 1968, 221.
15. **Crits, G. J. and McKeown, M. C.,** Ion exchange acid-alkali neutralizer for demineralizer, Proc. 22nd Int. Water Conf., Pittsburgh, PA, 1961.
16. **Kunin, R. and Vassilou, R.,** Regeneration of carboxylic cation-exchange resins with carbon dioxide, *Ind. Eng. Chem. Prod. Res. Dev.* 2, 1, 1963.
17. **Stewart, P. B.,** Seawater Demineralization by Ammonium Salts Ion Exchange, University of California Report Series No. 175, Issue 18, Berkeley, 1959.
18. **De Pree, D. O., Katzakian, A., and McGehee, D. C.,** Theory and Practice of Ion Exchange, Soc. Chem. Ind., Paper 28, London, 1976.

Chapter 9

WATER AS AN ELUTING AGENT FOR ION-EXCHANGE PROCESSES

C. Calmon

TABLE OF CONTENTS

I. INTRODUCTION

Among the unit processes utilized for recovering valuable products from waste, removing trace toxic metals, recycling of water, and other elements in pollution control, the ion exchange process is one of the most versatile. However, ion exchange per se can be a polluting process if the effluent from the elution or regeneration step requires chemicals which on passing through the bed cannot be utilized and must be discarded in streams, soils, or sewer lines. The volume of this effluent is smaller than the original volume of the waste solution and, therefore, can be discarded with better control.

Due to the polluting aspects of some ion-exchange processes, their use may be limited in areas where water sources are limited or brackish. A variety of processes have been developed to reduce the quantity of regenerant or eluting solution with the aim of attaining stoichiometry for the reaction. But, even if the stoichiometric dosage value is attained, the process still generates 100% more salts than exchanged, as illustrated by the water softening process. The softening step is

$$2\,RNa + Ca^{2+} \rightarrow R_2Ca + 2Na^+ \qquad (1)$$

For equivalent calcium removal, an equivalent amount of sodium is produced.

The regeneration step is

$$R_2Ca + 2Na^+ \rightarrow 2\,RNa + Ca^{2+} \qquad (2)$$

In this step, the calcium originally removed is recovered, and 100% more ions enter the water environment. On the average, however, most processes do not attain stoichiometry, so additional waste is produced. Of course, in systems where the eluting effluents can be reused, e.g., recovery of copper in the rayon industry, the problem of regenerant waste does not exist.

II. WATER AS AN ELUTING AGENT

Since water contains both H^+ and OH^- ions and since these ions are the most frequent regenerating or eluting ions, thought was given for many years to the possibility of using water as a regenerating solution. This was accomplished in the last 10 years. However, the concept of using water as an eluting solution in ion-exchange processes was first recognized by Moore and Kraus[1] of the Oak Ridge National Laboratory, Tennessee, who demonstrated it in two processes which will be described later. Today, water as an eluant has been utilized in reactions involving:

1. A change in an ionic complex in the presence of water
2. A change in valence of an ion on contact with a dilute solution, i.e., water
3. A Donnan membrane equilibrium
4. Special exchangers
   a. The affinity of the exchanger for certain ions varies with the temperature.
   b. The affinity of the exchanger for ionic constituents are greater than for nonionic entities.

A. Change in Ionic Complex in Water

Two industrial processes are operating today in which water is used as an eluant involving ions to transform an ionic complex to a normal salt.

1. Removal of Iron in Concentrated HCl

Moore and Kraus[1] noted that $FeCl_3$ in HCl forms a complex ion, $FeCl_4^-$, which is readily removed from solution by an anion exchanger. When the complex is in contact with water, it dissociates to $FeCl_3 + Cl^-$ so the anion exchanger releases $FeCl_3$ and is converted to the chloride form, as can be seen from the following equations:

$$FeCl_3 + Cl^- \rightarrow FeCl_4^- \qquad (3)$$

$$RCl + FeCl_4^- \text{ (in HCl)} \rightarrow RFeCl_4^- + Cl^- \qquad (4)$$

$$RFeCl_4^- \text{ (in water)} \rightarrow RCl + FeCl_3 \qquad (5)$$

This process is applied to the purification of hydrochloric acid for upgrading.

2. Zinc Recovery from Pickle Liquor

Another industrial process involving a complex is the recovery of zinc from pickle liquors. This process was developed at the South African National Institute for Metallurgy by Haines and co-workers.[2,3] Spent pickle liquor is used for stripping zinc from rejected galvanized products and from accumulation on jigs during galvanizing operations. This mixture is passed through an anion exchanger where the zinc chloride is picked up as the $ZnCl_4^{2-}$ complex, while the effluent containing $FeCl_2$, which does not form a complex in this solution, and residual HCl is mixed with pickle liquor free of zinc. This mixture is sent to a spray roaster where the iron is recovered as iron oxide and the chloride as HCl. The former is sold as a low grade pigment, while the latter is recovered to a strength sufficient to be used for pickling steel.

The anion exchanger is treated with water for recovering the zinc as zinc chloride. It then goes through a liquid cation-exchange extraction from which the zinc is recovered as zinc sulfate and from which the zinc is recovered electrolytically. This "Metsep" process, as it is called, forms a closed system for recovery of the iron as the oxide, the Cl as the acid, and the zinc as the sulfate from which the zinc can be recovered.

B. Change in the Anion Valence

Sulfuric acid in solution exists as H^+ and HSO_4^-. In very dilute solutions or in water, HSO_4^- goes to H^+ and $SO_4^=$. Thus, in acid, an anion exchanger picks up the sulfate as HSO_4^- which occupies one polar group, i.e.,

$$R_2SO_4 + HSO_4^- + H^+ \rightarrow 2RHSO_4 \qquad (6)$$

However, when water is passed through the anion-exchange bed, the HSO_4^- is converted to $SO_4^=$, which occupies two positions, while one $SO_4^=$ is released as can be noted from Equation 7:

$$2RHSO_4 \rightarrow R_2SO_4 + H_2SO_4 \qquad (7)$$

This observation by Kraus et al.[4] led Odland and Pabich[5,6] to apply it as a desalination process especially for waters containing sulfate ions. Water containing sulfate salts is passed through a cation exchanger in the H^+ form (i.e., regenerated with acid), and the salts are converted to the corresponding acid,

$$\begin{array}{l} 2\,RH + CaSO_4 \rightarrow R_2Ca + H_2SO_4 \\ \qquad\qquad\qquad\qquad\qquad \downarrow \\ \qquad\qquad\qquad\qquad\quad H^+ + HSO_4^- \end{array} \qquad (8)$$

The effluent is passed through an anion exchanger in the $SO_4^=$ form with desalination following as in Equation 6 and elution being carried out as in Equation 7.

The process of desalinating brackish water by this procedure is referred to as the Sul-bi-Sul® process. Variations in the process depend on the water composition. Details of the process are given in References 5 and 6.

C. Utilization of Donnan Potential

1. Ion Exclusion

An ion exchanger can be considered to be a concentrated ionic solution with a fixed charge on the matrix of the polymer and a free counter-ion. Ions in solution (with which an exchange particle is surrounded), as well as the counter-ion of the exchanger, are free to diffuse in and out of the particle. The result is that since electroneutrality must exist in both the exchange particle and solution a potential develops on the particle to prevent a buildup of mobile ions from rising above an equilibrium value. The concentration of the electrolyte in solution is greater than in the exchange particle, i.e., the free electrolyte is partially excluded by the ion exchanger. Since no such restrictions exist for organics or weak electrolytes, separation of the latter from strong electrolytes is possible. The technique of separating ionic components from nonionics by this exclusion was developed by Bauman and co-workers of Dow Chemical Company.[7,8] Thus, when a solution containing hydrochloric acid and ethylene glycol is passed through a cation exchanger, the first compound to appear in the effluent is hydrochloric acid. At breakthrough of the organics, the service run is stopped. The elution of the ethylene glycol from the exchanger is done with water. Many similar separations can be accomplished.

There are many factors on which the efficiency of the process depends, such as type, capacity and cross-linking of the exchanger, solution concentration, size of ions, bed dimensions, flow rate, temperature, and mode of operation. The only chemical used in the fractionation or separation of ionic from nonionic species is water.

2. Acid Retardation

This process has been applied to separate mineral acids from their salts through differences of distribution with an exchanger as a result of differences in the degree of ionization of these electrolytes in the solution. A good example of this is the separation of residual phosphoric acid from aluminum phosphate in a solution produced in the Bright-Dip Process. A solution containing 10% aluminum phosphate and 30% phosphoric acid was separated into its respective electrolytes by passing the solution through a strongly basic anion exchanger in the phosphate form.[9] The phosphoric acid retained by the exchanger was eluted with water. Keller and Morrison also applied this process to separating nitric acid from ferric nitrate.[9] Many of the limitations in the process were eliminated by the use of a special moving-bed contactor.[10]

D. Special Ion Exchangers

1. Sirotherm Process

This process was developed at the Chemical Technology Division of CSIRO, Melbourne, Australia, by Weiss and co-workers.[11,12] The affinity of the exchangers for given ions depends on the solution temperature. The exchangers have the unique feature that in the salt form they regenerate with hot water and not with chemicals; the advantage is that there is a reduction of the salts discharged back to the environment.

The ion exchangers are of the weak-acid or weak-base types which when combined in a bed can be used for desalination similar to that obtainable with standard ion exchangers. The process, however, is particularly applicable to brackish waters (1000 to 3000 ppm), preferably those containing sodium salts. The salt content can readily be reduced to the range of potable water, namely, about 500 ppm.

Brackish water at ambient temperature (≈20°C) is passed through a column of these special exchangers. The salts are exchanged for H^+ and OH^- ions which form water. The leakage is high; on breakthrough, the resins are regenerated with hot water (heated with waste heat) at 80°C. About 15% of the raw water is required for the regeneration.

There are limitations to the process; the water has to be low in hardness, pH control is essential, rate of heat transfer, etc. Many of these problems have been overcome and large plants are now in construction. Also, improved ion exchangers are being developed.

2. Ion Retardation

Ion exclusion is attained through an ion exchanger that contains only one kind of a polar group. However, when both cationic and anionic groups are present in the exchanger in very close proximity, the exchanger tends to permit the diffusion into its framework of ions from solution while limiting the diffusion of organics or weakly dissociated electrolytes. Thus, it is possible to separate ionic entities from nonionics similar to the ion-exclusion process, except that, in the ion-retardation process, the first effluent from the column contains the nonionic and the water eluent contains the ionic components. This process was developed by Hatch and co-workers of Dow Chemical Company, Midland, Mich.[13,14]

III. CONCLUSIONS

Ion exchange offers many possibilities for utilizing water as an eluant through complex ion formation, novel ion exchangers, a change in valence, and charge potentials which are built up on the exchangers. In view of the present emphasis of reducing the degree of pollution to the environment, analyses should be made in plants to note whether any of the above are applicable to plant processes with a view towards economic gain through recovery and purification of products as well as from reduced quantities of wastes. It is to be noted that half of the processes mentioned were developed through government research in Australia, the U.S., and South Africa.

REFERENCES

1. **Moore, G. E. and Kraus, K. A.**, Adsorption of iron by anion exchange resins from hydrochloric acid solutions, *J. Am. Chem. Soc.*, 72, 5792, 1950.
2. **Haines, A. K., Tunley, T. H., TeRiele, W. A. M., Cloete, F. L. D., and Sampson, T. D.**, The recovery of zinc from pickle liquors by ion exchange, *J. S. Afr. Inst. Min. Metall.*, 74, 149, 1973.
3. **Tunley, T. H., and Sampson, T. D.**, The Metsep process for separation and recovery of zinc, iron and hydrochloric acid from spent pickle liquor, *J. S. Afr. Inst. Min. Metall.*, 77, 243, 1976.
4. **Kraus, K. A., Nelson, F., and Baxter, J. F.**, Separation of sulfuric acid from metal sulfates by ion exchange, *J. Am. Chem. Soc.*, 75, 2768, 1953.
5. **Odland, K. and Pabich, H. L.**, Desalination by the Sul-bi-Sul process, Proc. 26th Int. Water Conf., Pittsburgh, 1965.
6. **Odland, K. and Pabich, H. L.**, Demineralization of brackish water, *Ind. Water Eng.*, December 1965.
7. **Bauman, W. C.**, U.S. Patent 2,684,331, 1954.
8. **Bauman, W. C., Wheaton, R. M., and Simpson, D. W.**, Ion exclusion, in *Ion Exchange Technology*, Nachod, F. C. and Schubert, J., Eds., Academic Press, New York, 1956, 182.
9. **Keller, H. W. and Morrison, W. S.**, Acid Recovery from Stainless Steel Pickling Wastes by Acid Retardation, 148th Meeting of American Chemical Society, 1964.
10. **Lancey, L. E., Stewart, F. A., and West, J. N.**, Pilot Plant Optimization of Phosphoric Acid Recovery, Project S-802637, Environmental Protection Agency, Cincinnati, 1974.
11. **Weiss, D. E., et al.**, The Sirotherm demineralization process-anion exchange process with thermal regeneration, Aust. Chem. Inst., 1965.
12. **Bolto, B. A., Weiss, D. E., Cope, A. F. G., and Stephens, G. K.**, in *The Theory and Practice of Ion Exchange*, Streat, M., Ed., Society Chemical Industry, 1976.
13. **Hatch, M. J., Dillon, J. A., and Smith, H. B.**, Preparation and use of snake-cage polyelectrolytes, *Ind. Eng. Chem.*, 49, 1812, 1957.
14. **Hatch, M. J. and Dillon, J. A.**, Acid retardation, *Ind. Eng. Chem. Process Des. Dev.*, 2, 253, 1963.

Chapter 10

LABORATORY EVALUATIONS OF ION-EXCHANGE PROCESSES

S. A. Fisher and F. X. McGarvey

TABLE OF CONTENTS

I. INTRODUCTION

The first question that a laboratory study of an ion-exchange pollution control process should answer is whether or not ion exchange should be a component of the system at all. This question is sometimes answered without taking a bead from a bottle — by a thoughtful consideration of the nature of the stream to be treated, the ions present in it and their concentrations, the nature of the effluent to be produced, alternative processes that could be used, and the economics of operating the system.

Ion exchange is a process most economically suited to the treatment of large volumes of solution containing low concentrations of the ions to be removed where an effluent substantially free of one specific ion, or type of ion, is desired. Where concentrated streams are encountered, it may be only a final step in a purification process used to polish the product from other treatment processes, to remove and concentrate a specific valuable or toxic component, or to render the stream suitable for reuse. In general, the product from an ion-exchange process is too good to be thrown away because of its economic value. The possibility of the reuse of the treated stream either as process water itself or, in the case of partial deionization, as chemical make-up, should always be a part of the thinking in the design of a laboratory feasibility study of an ion-exchange system.

If ion exchange is to be considered as a viable alternative for a potential pollution control process, what information should be sought in a laboratory-scale study? The following is a partial list of questions that should be asked in a laboratory-scale process development program:

1. What resin or resins should be used?
2. What volume of solution can be treated per volume of resin?
3. What kind of regenerant should be used at what concentration and in what amount?
4. What should be the quality of the effluent stream?
5. What amount and concentration of waste will be produced during regeneration and how can it be disposed of?
6. Is the process repeatable so that the resin can be used for many cycles?
7. Are there parameters that cannot be studied on a laboratory scale but must be studied on a pilot or larger-scale operation?
8. What are the projected operating costs?

Careful design of this program is then of the utmost importance so that the maximum amount of information may be squeezed out of the minimum amount of experimental work. However, before an experimental program can be designed to study a potential ion-exchange process, experimental work is required to provide an accurate and complete knowledge of the nature of the stream to be processed. Without this preliminary work, laboratory-scale studies can at best yield approximations of reality which may be of little value.

II. ANALYSIS OF PROCESS STREAMS

To conduct a laboratory-scale ion-exchange process study without a reliable and complete analysis of the stream (or streams) to be treated and a knowledge and understanding of the process from which the streams come is usually a total waste of time. Too often, particularly in pollution analysis, streams are analyzed only in terms of discharge permit limits, and high concentrations of ions not included in the permit may be present but unlisted on the analysis sheet. Knowledge of what these are and their concentrations is essential in the study of an ion-exchange process.

Just as important as the analysis of the sample itself is the assurance that the sample analyzed has been properly obtained. Where a stream from a existing process is to be treated, this is merely a question of proper sampling and compositing. However, if the ion-exchange process is to be fitted at the end of a chain of treatment processes not presently in existence, it is sometimes necessary to miniaturize the entire process first to produce samples for analysis and then the influent to be used in the test work itself.

Such work on the analysis of the system may even require analytical work on the chemicals being used in the process. In metalworking systems, for example, proprietary mixtures are frequently used in the baths. Some of these may contain additives which, even if present in small amounts, will influence an ion-exchange system by fouling resins, by precipitating in the bed as the pH changes, or by being totally oblivious to the treatment process and accumulating in the water intended for recycle. Analysis of the concentrates added to the bath is sometimes the quickest way to find out what to look for in the process stream, where these minor additives may be present only in very low concentrations.

In any case, it is important that the analysis of the influent for a laboratory study of an ion-exchange process includes the measurement of total dissolved solids as well as the measurement of a sufficient number of individual components to account for these dissolved solids and establish a reliable anion-cation balance. Such an analysis is frequently implemented by the use of ion exchange itself to remove all cations by a hydrogen form resin. The titration profile of the acids so produced is used not only in calculating the total ionic strength of the solution but also for the study of the distribution of acid strengths produced from the anionic components. This information is a valuable aid in the selection of the anion-exchange resins to be used in the process.

III. PROCESS SELECTION

Armed with a reliable analysis of the solution to be treated by ion exchange and an understanding of the process from which the waste comes, it is finally possible to decide first, if ion exchange should be used at all, and, if so, what kind of an ion-exchange process seems most practical for the specific situation. It is possible just to use a cation resin and return its effluent, now containing only acids, to the process as make-up water? Would a softening procedure for removal of toxic or otherwise objectionable cations with reuse of the product water or discharge of the treated stream, now containing only sodium salts, be possible? Should the stream be totally deionzed? What values are recoverable? With the knowledge at hand, it is now possible to project the feasibility of these alternatives and to select from them a process to be investigated in the laboratory.

IV. RESIN SELECTION

Usually, the first step in a process design study is the selection of the resins to be used. Sometimes this choice is made obvious by the process requirements such as operation at an elevated temperature or a high flow rate. Sometimes the choice is based on the known selectivity of a particular resin for an ion to be recovered in the process. Conversely, the selection may be based on a given resin not picking up an ion that can safely be passed into the treated stream. Standard deionization processes using combinations of strong-acid-weak-base resins or strong-acid-strong-base resins involving only ions normally found in natural water supplies can usually be designed from data found in the manufacturer's technical literature without laboratory work beyond the analysis of the stream to be treated. Hence, it is only to the nonstandard process that our attention is directed here.

Even in the case of a nonstandard process, the resin choice will usually be sharply limited by the intent of the process. Unless the process involves unusually concentrated solutions, removal of large complex ions, recovery of organic compounds, or selective recovery of a single ion, it is usually possible to narrow the possible resin choices down to a single anion-type (e.g., quaternary or weak-base) and a single cation-type (e.g., sulfonic or carboxylic) on the basis of the analysis of the solution and process requirements without laboratory work at all. Resin selection experiments in these cases may be limited to the comparison of the operational efficiency of two resins of the same functionality but different porosity or polymer backbone, as say, a gel or a macroporous sulfonic resin. Comparison of a Type I (trimethyl amine) and Type II (dimethylethanol amine) quaternary or a methacrylic and an acrylic acid carboxylic acid resin may also be involved. Only in the case where weak-base resins are to be employed is the variety of polymer and functionality types large enough and the performance characteristics variable enough to warrant a major screening program before a single resin, or perhaps two or three resins, are selected for detailed performance studies.

V. BATCH AND COLUMN STUDIES

Two modes of conducting initial resin screening tests are open to the investigator: batch equilibrium studies and column tests. For ultimate design purposes, there is no question that column runs are an essential part of the laboratory study. The question then is "Are batch studies of any use at all?"

By batch tests we mean a system wherein a known volume of resin is contacted with a known volume of solution and the change in concentration of the solution phase is measured either as a function of time or after a set contact period. Such experiments are quite useful when the large number of types of resin are to be screened for the removal of a specific solution component, where large molecules of unknown resin affinity are involved, or where a number of possible regenerants and regenerant concentrations are to be evaluated. In these cases, batch tests can be extremely useful in quickly narrowing down the possibilities to the most fruitful area for further work. They are also extremely valuable when the effect of temperature on either the loading or elution cycle is to be evaluated. Since relatively small amounts of feed solution are required, such tests are also of value when the supply of solution to be treated is limited.

A. Batch Equilibrium Tests

As noted above, the usefulness of batch tests is in the preliminary testing period where they are used to narrow down variables and to select the most promising areas for detailed study. The specific details of the test will vary widely depending upon the type of information being sought, solution concentrations, and the volumes required in the analytical procedures used to follow the changes occurring. Where resin types or ionic forms are being compared, a set volume of solution may be equilibrated with varying known weights or volumes of resin and the depletion of the liquid phase measured and plotted as shown in Figure 1. Such curves, which show that resin B has the best affinity for the ion of interest, must be combined with desorption curves such as shown in Figure 2. In the latter case, the curves show that resin B is equally reluctant to release the material to the regenerant. If the ultimate process here is to be a cyclic one where the ion of interest is to be recovered by elution, or at least the resin phase is to be regenerated for repeated use, a less selective resin such as resin C will be the better choice. Desorption curves of this type would normally be run with the resin, which is loaded to capacity, being equilibrated with known volumes of regenerant.

Similar curves can be drawn to show the effects of rate and temperature. However, it should be noted that these experiments all depend on what happens to one or at most two or three ions in a given system. In a complex waste, the value of the results obtained diminishes rapidly in comparison with the results from column runs.

B. Column Tests

1. Column Size

The question often asked by laboratory investigators is "How small must the column be to generate data which is useful in the design of real plants?" The real answer to that question involves the rapport between the chemist running the laboratory columns and the engineer designing the system and their individual competencies and understanding of the ion-exchange process.

Industry is in agreement that results from experiments in a 25-mm(1-in.)-diameter column can be scaled-up with confidence. A column of such dimensions, similar to that shown in Figure 3, is specified in standard performance test procedures by ASTM[1] that are used routinely in writing specifications for materials to be used in conventional water treatment processes. Furthermore, the data supplied for design calculations by the resin manufacturers has in most cases been derived from experiments on this scale.

To fill a 25-mm-diameter column to the required 750-mm minimum depth, however, requires a resin volume of nearly 375 mℓ. Let us suppose that the resin is a typical sulfonic resin with a capacity of 1.8 meq/mℓ. Such a column then has a potential

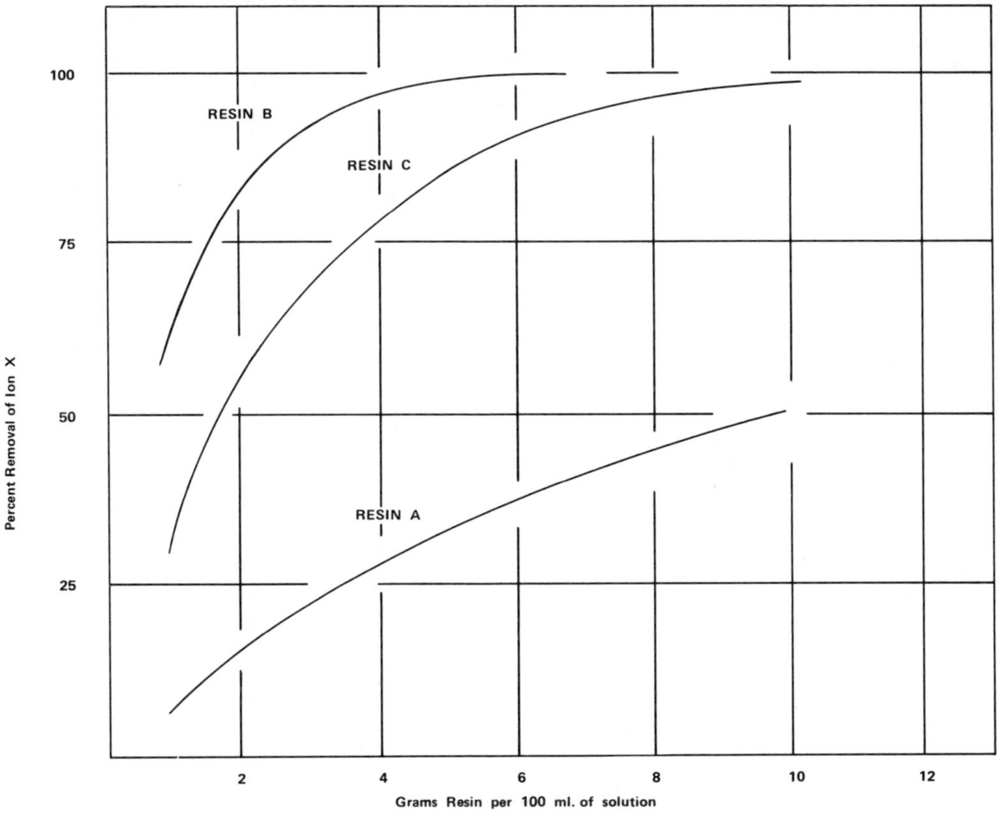

FIGURE 1. Batch test data. Depletion of the solution phase as a function of resin weight.

capacity of 0.675 eq. Now if the concentration of the ion to be removed is 5 ppm, which for a typical cation will represent a concentration in the range of 10^{-4} eq/ℓ, a single run might require 6750 ℓ or 1780 gal of influent. This is hardly a laboratory-scale experiment even if the working capacity were only 10% of theoretical.

Obviously then, smaller columns must be considered when very dilute solutions are concerned and where only limited amounts of solution are available. Old-timers in the field will cite the example that the 200 ft^3 units used for uranium recovery from South Africa gold tailings were designed directly from data obtained with 25 mℓ of resin contained in 50-mℓ burettes of approximately 10-mm diameter. Careful work in columns of that size can be scaled-up to determine throughput capacity and regeneration efficiency. Even smaller columns with 5-mm diameters have been used where the amount of influent solution available is very limited. Work at this level, however, should be primarily aimed at making the process feasibility decision.

As noted above, apparatus as simple as titration burettes may be used for small-scale work. Modular columns intended for chromatographic work are also very convenient for laboratory-scale ion-exchange work. A wide range of diameters and column heights are available with connectors that permit the interchange of tips and tops and the connection of several columns in series. These also come with several choices of supporting plates. Polyester screen cloth cemented between rubber gaskets with liquid rubber also serves as a durable inert support that has the advantage of not plugging with colloidal material.

Regulation of flow rate and collection of precise effluent volumes is the greatest problem in operating small columns. Since many resins shrink or swell in changing

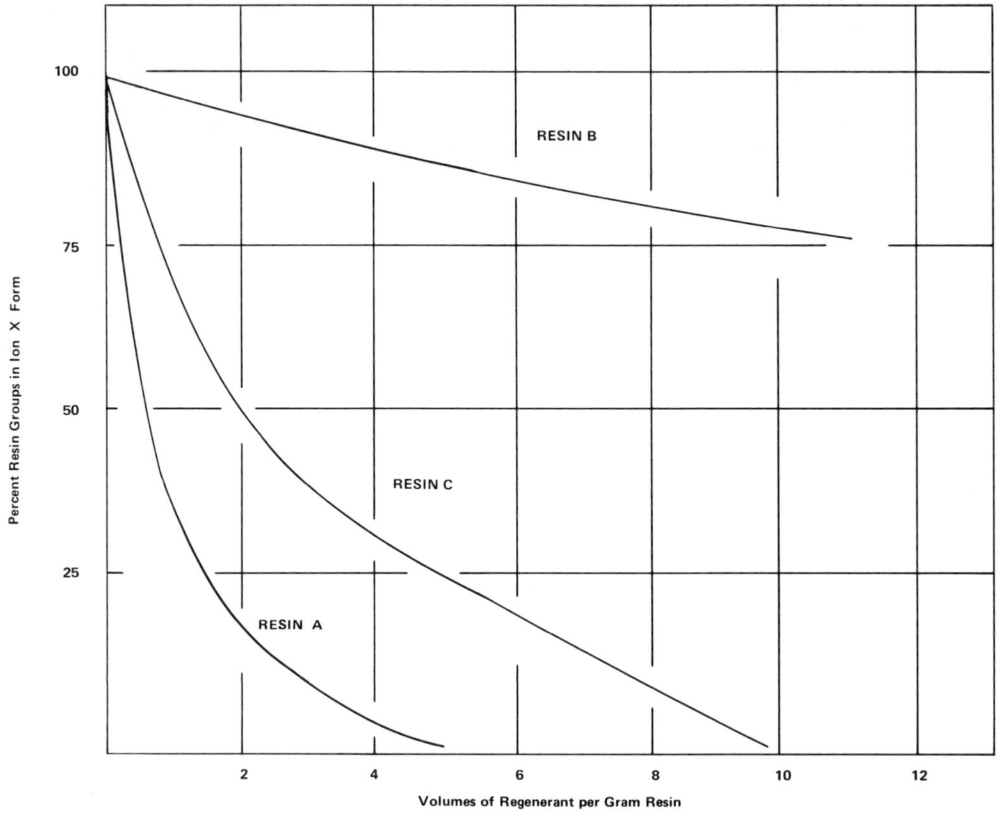

FIGURE 2. Batch test data. Removal of ions from resin phase by regenerant solution.

from one ionic form to another, the pressure drop through the bed may also change during a given run. This results in changes in flow rate as the resin volume changes. Hence, collection on a time sequence must be combined with a flow system assuring constant delivery if columns are monitored automatically. With care, all of the normal operations of a fixed-bed unit can be performed in these small beds except operations involving the separation and regeneration of two resin components such as in a mixed-bed or a bed containing two anion or two cation resins. In such cases at least a 25-mm (1-in)-diameter column should be used.

2. Test Variables

A number of test variables should be studied in the laboratory-scale system. In addition to resin types already mentioned, flow rate, temperature, concentration, quantity, and kind of regenerant, and co-current vs. counter-current flow may be studied. Which ones are selected in a given case is frequently dictated by factors other than the ion-exchange process itself, such as temperature requirements in the system, space limitations, and regenerant disposal problems. However, any or all of those listed can be studied in a well-designed laboratory-scale process study.

3. Initial Experiments

A process study has to begin somewhere. It seems logical to use in the first experiments the most promising combination of resins and operating conditions that can be selected based on previous personal experience and the technical literature. In pollution control systems, where the waste streams are highly variable, it is unlikely that anyone

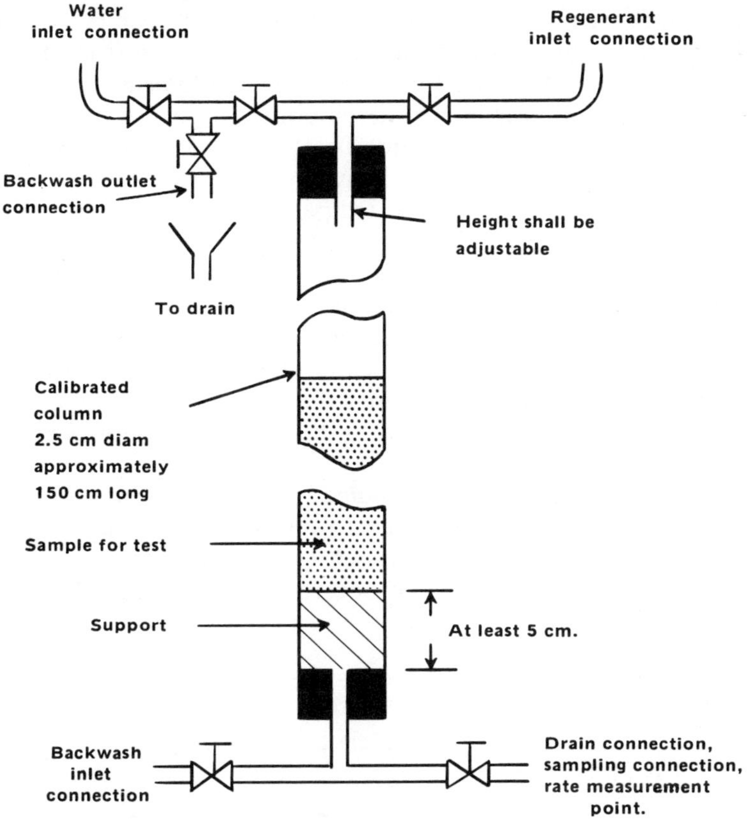

FIGURE 3. Typical column for resin tests. (From *Annual Book of ASTM Standards*, Part 31, Water Section X, Water Treatment Materials, American Society for Testing and Materials, Philadelphia, 1978. With permission.)

has studied and published information on precisely the system of interest. In the absence of specific information, the operating conditions with regard to flow rate and regenerant concentrations and amounts that are normally used in water treatment operations are often a good starting point. Literature available from resin manufacturers all suggest types, concentrations, and amounts of regenerants to be used for specific resins. In addition, standard performance tests available from the ASTM[1] provide specific operating instructions for testing the performance of different classes of resins in the laboratory. These are based on practical commercial experience and may be used as a starting point for a process study.

4. Test Measurements

Once the initial experiments are completed, changes in variables can easily be evaluated by comparing the new performance with the base run. Certain effects are, of course, predictable such as:

1. Effluent concentration (leakage) increases with increased flow rate during the operating cycle.
2. Effluent concentration decreases (or remains constant) with increased volume of regenerant.

The effects of other variables such as the type and concentration of the regenerant and solution temperature are not always so predictable and may require careful study. This is particularly true when materials of value are to be recovered from the regenerant or its volume must be minimized because of waste disposal problems.

The specific measurements made in the laboratory in a process study depend, of course, on what the process intends to do. Total deionization processes are most frequently monitored by following the conductivity (or resistivity) of the effluent stream. Here the volume of waste that can be treated per volume of resin as a function of the amount of regenerant is most frequently determined. Where one single ion in the stream is of interest, it may be monitored, and the volume of solution that may be treated before it reaches a specific level in the effluent is determined. Where materials of economic value are to be recovered, multiple columns in series may be studied. Here the volume of solution required to saturate one column with a specific ion of interest may be the parameter measured. Sometimes the degree of saturation of the first column before leakage reaches a preset value on the second, third, or even fourth column in the series is determined.

Data from such experiments may be handled in a number of ways. Frequently, plots of volume of solution treated per volume of resin (commonly called bed volumes) as a function of the amount of regenerant used are drawn as shown in Figure 4. Leakage values for a specific ion as a function of bed volumes treated may also be plotted as

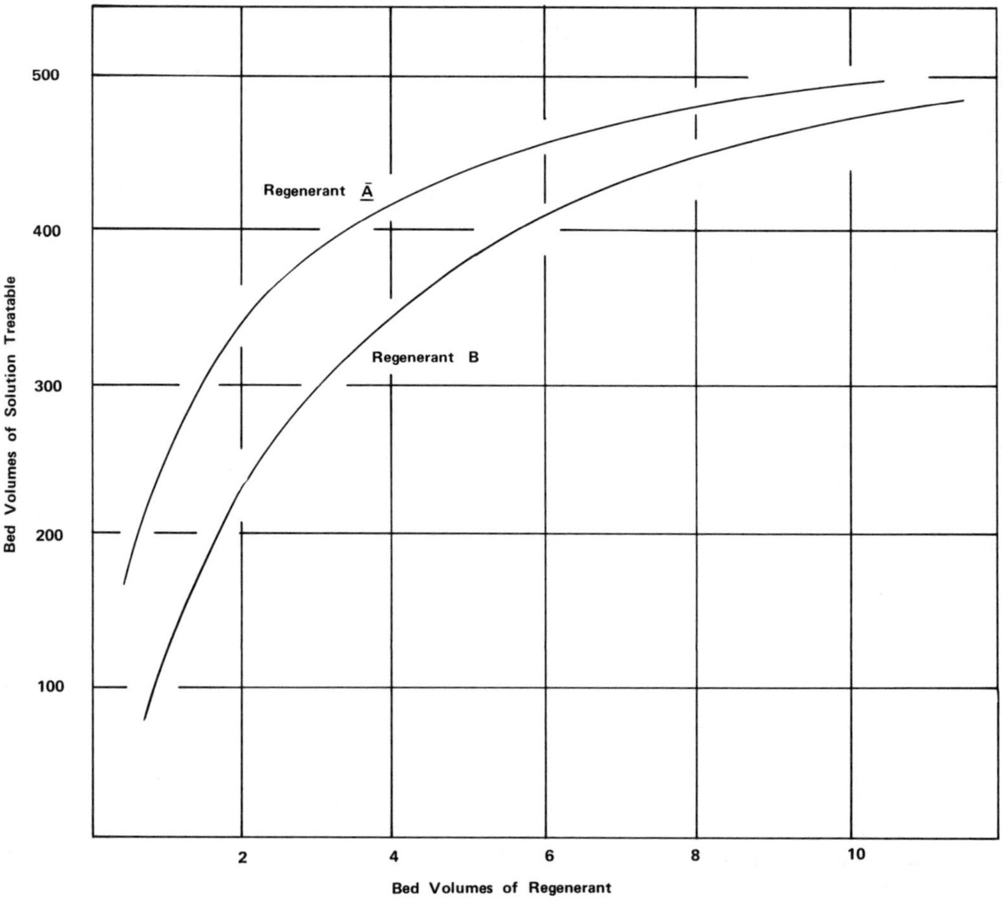

FIGURE 4. Column performance tests. Volumes of solution treated per volumn of resin as a function of regenerant used.

shown in Figure 5, where typical leakage curves from a three column operation are shown.

On the regeneration cycle, it is sometimes of value to determine salt profiles in the effluent as well as excess acid, as shown in Figure 6. Such curves may be used to split a regenerant stream, in this case perhaps at 1.25 bed volumes, discharging the first portion as waste and reserving the remainder for reuse, with make-up acid, in the next cycle. Similar curves are constructed when materials of economic value are to be recovered so that cuts with maximum concentration of the ion of interest may be segregated for the recovery process and tailings recycled as regenerant in the next cycle as shown in Figure 7.

5. Long-term Resin Performance

The most difficult part of ion-exchange operation to predict from laboratory-scale tests is resin life. Where resins are used to treat wastes containing weakly ionic materials that may foul the ionic sites and oxidizing or reducing agents the question frequently asked is "How many cycles can I expect from a resin charge?"

Such resin life evaluations involve repeated, frequently short, cycles of loading and regeneration followed by a reevaluation of the cycled resin. The reevaluation may be simply a redetermination of basic resin properties such as available exchange sites or may involve a complete rerun of a performance test. Such short-cycle tests have long established their value in selecting between alternative resin candidates for a given application. Extrapolation to absolute numbers in terms of resin life is done with variable success, since in some situations resin degradation will be dependent on total contact time and hence cannot be predicted from short-term tests.

The prediction of physical life on the basis of laboratory tests is also an uncertain venture. Where resins break up physically due to osmotic shock, in cases where they are cycled from solutions of high concentration to essentially pure water, comparative cycle tests in the laboratory will serve to screen out the best candidates. Breakdown

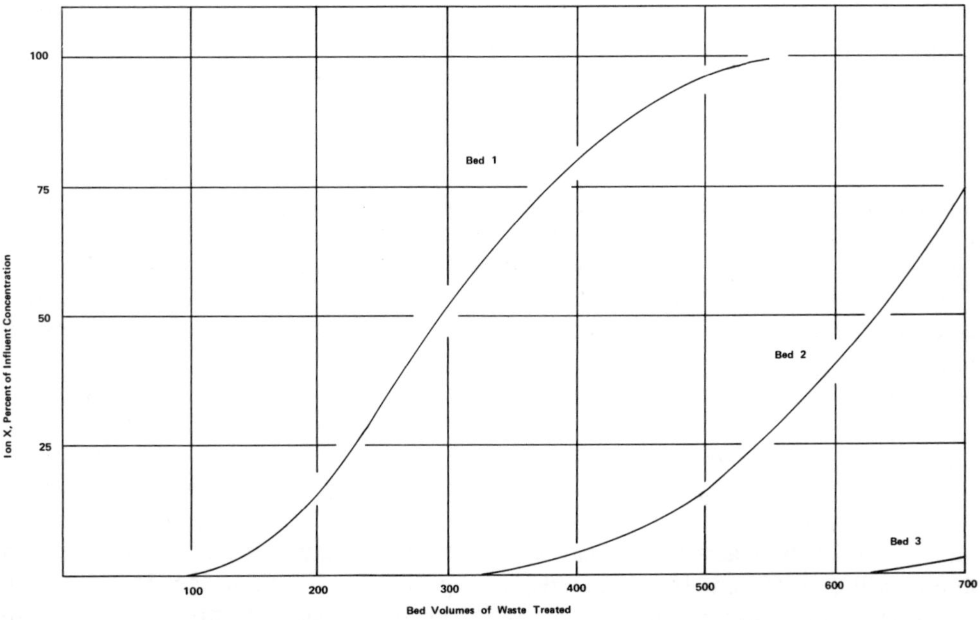

FIGURE 5. Column performance tests. Leakage data from three columns operated in series.

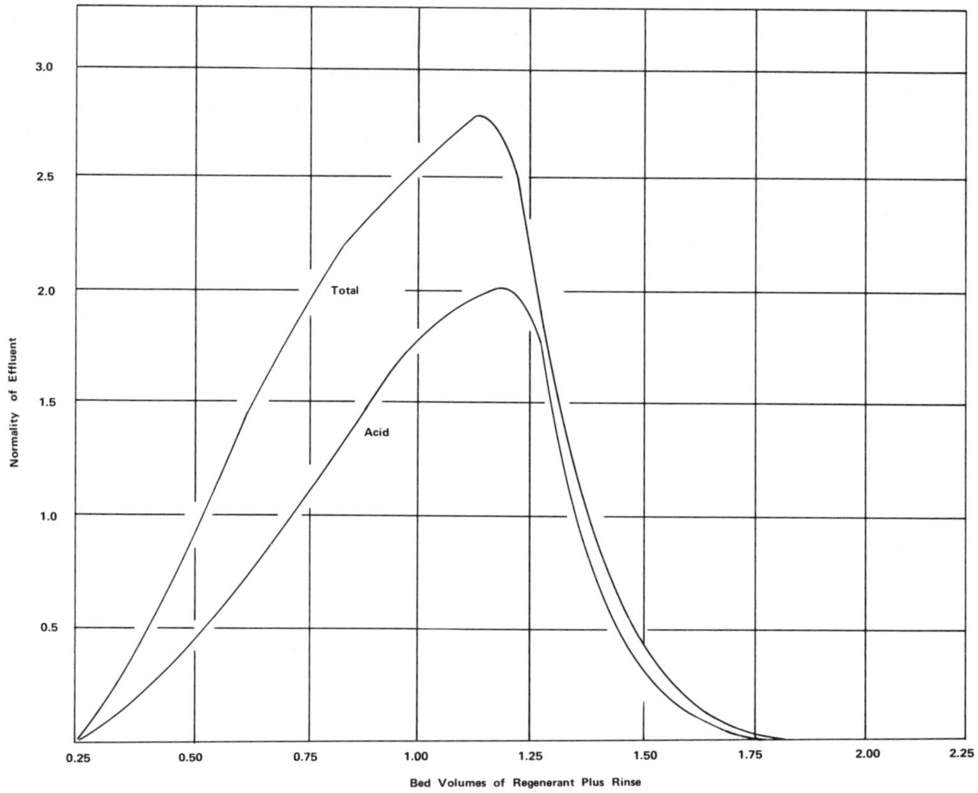

FIGURE 6. Regeneration profile data. Total salt content and residual acid in effluent during regeneration.

due to changes in applied pressure in the operation of deep beds or to attrition during resin transfer operations is much more difficult to simulate on a laboratory scale. In such cases, laboratory tests should, if possible, include resins of already demonstrated performance in process applications so that other candidates may be compared to them.

VI. LIMITATIONS AND PITFALLS

Ion exchange is a sophisticated chemical process. It is not possible in a few pages to touch upon all of the possibilities and all of the problems that may be encountered in a laboratory-scale test program or in the operation of a plant. Let it suffice to say that it is wiser to detect problems in a laboratory-scale operation and overcome them, or abandon the project, than to try to alter a plant already built. To this end a few generalized warnings are appropriate here.

Perhaps the greatest danger in a laboratory study is to fail to replicate the data. Ion exchange is basically an equilibrium process, and results on new resins are never typical of the operation to be expected in a repetitive process. The main reason for this, of course, is that resins usually come from the manufacturer in an ionic form that is different from the ions in the process stream. Until the column has been equilibrated with the ions in the waste to be treated, the results, in terms of leakage and treatable volume, will be different from those to be expected in routine usage. In some cases, depending on the ions originally in the resin, performance will improve on repetitive cycling. In others it may get worse. In any case, it is dangerous to base a process design

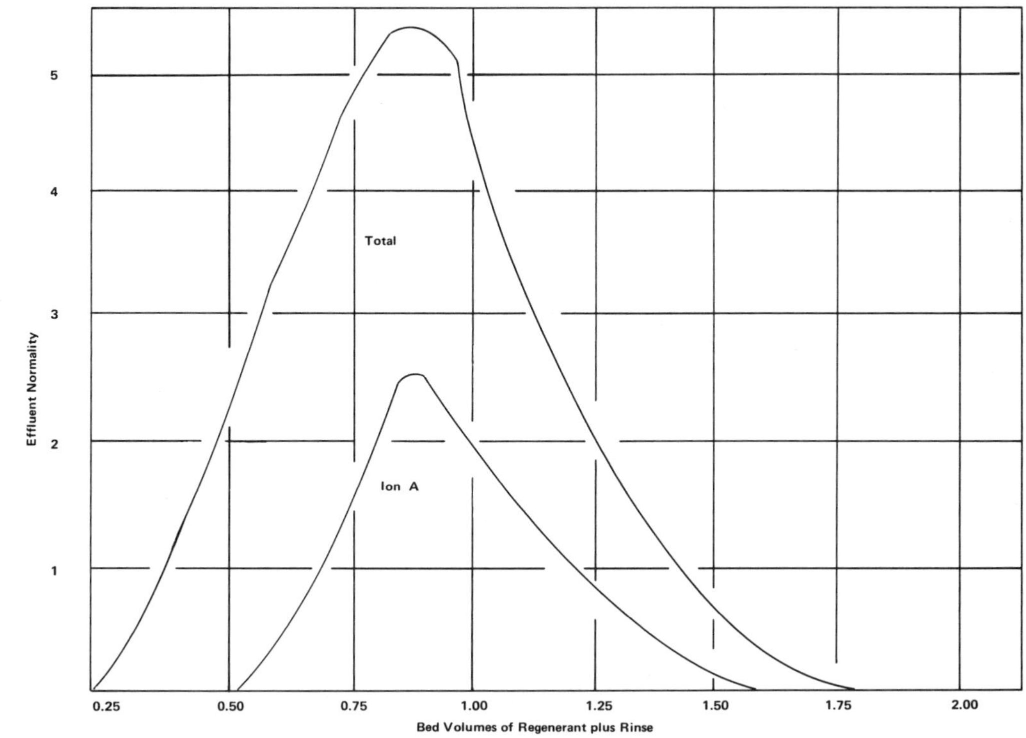

FIGURE 7. Regeneration profile data. Concentration of an ion of interest in regenerant solution.

on data that does not represent at least two duplicate performance runs; three successive runs producing similar data greatly increase the confidence level.

Another potential pitfall in the scale-up of a laboratory process is to fail to appreciate that the volume that a resin occupies is a function of its ionic form. In a small-scale experiment, changes in volume may be so small that they go unnoticed. Failure to take these into account in a full-scale plant may result in misplaced distributors or beds with inadequate backwash room. Particular care in this respect should be taken when polymers with weak-acid or weak-base functionality are employed since volume changes of the order of 100% are sometimes encountered in these systems. However, even in the commonly used sulfonic cation and quaternary ammonium anion resins, volume changes are encountered, particularly in the polymers of lower cross-linkage.

Even more important than anything else is the nature of the solution used in the laboratory work. The fact that the entire validity of a laboratory study evolves from the validity of the solutions used in the study bears repeating. Real waste should be used if available. Where it is not, laboratory studies of an ion-exchange process should be made with solutions containing all the anticipated components. To do otherwise is to court trouble.

REFERENCE

1 Annual Book of ASTM Standards, Part 31, Water Section X, Water Treatment Materials, American Society for Testing and Materials, Philadelphia, 1978.

Chapter 11

ION EXCHANGE AS AN ANALYTICAL TOOL IN POLLUTION CONTROL

H. F. Walton

TABLE OF CONTENTS

I. INTRODUCTION

Ion exchange is used to detect and measure dissolved pollutants in water. It is used in two ways: first, to collect and concentrate traces of impurities from large volumes of water, and second, to separate various substances of interest after they have been concentrated. Both inorganic and organic contaminants may be collected.

The analysis of trace contaminants in water is a very active field today. Ion exchange is playing a part, along with absorption by non-ionic polymers.

II. PRECONCENTRATION BY ION-EXCHANGE RESINS

A. Inorganic Ions

Ion exchange is a competitive process, and it is effective for preconcentration only if the exchanger is selective for the ions of interest. Fortunately, in pollution control, these are primarily the "heavy metal" ions, and they are strongly held by most resins. They are especially strongly held by chelating resins like Chelex-100® and by resins with functional carboxyl groups. The resins can be used in columns or by stirring and batch equilibration.

For use in columns, it is best to have chelating and carboxylic resins in their sodium form, which is ionized and swollen, rather than the hydrogen form, which is non-ionized, non-swollen, and reacts slowly. A large volume of the water to be analyzed is passed through a small column of resin, then the absorbed metal ions are stripped from the column by a small volume of acid, or (for metals like Mo or V) aqueous ammonia. The concentrations in the eluate are measured by a suitable technique, like atomic absorption or anodic stripping voltammetry.[1]

Batch equilibration has the advantage that less resin is required than in columns.

Typically, 200 mg of powdered resin is shaken for a few hours with 100 to 200 mℓ of solution. Then the metal ions, instead of being extracted with acid, are determined in the resin itself by compressing the resin into a pellet and using X-ray fluorescence.[2-4]

A novel technique is to suspend a strip of cation-exchanging membrane in a flowing river for 3 to 4 days, in the course of which the membrane takes up heavy metal ions and comes to equilibrium. The final metal loading depends on the initial concentrations in the river water, which may be as low as 10^{-7} *M*, and also on the concentrations of major ions such as calcium and sodium. The metals are identified and their amounts measured by X-rays produced by proton bombardment.[5]

Anion-exchange resins absorb negatively charged metal complexes, such as chloride complexes, in a highly selective way. Several studies[6,7] describe the use of these resins in columns to collect trace metals like Pb, Cu, Cd, Zn, Mo, and U from natural waters. The metals are stripped from the columns by reagents like 6 *M* hydrochloric acid, then measured by atomic absorption or a color-producing reagent like dithizone. Selective absorption is achieved by adding an appropriate complex-forming reagent to the water before it is passed through the column; for example, Pb, Cd, and Bi are absorbed by a strong-base anion exchange resin from a solution 0.15 *M* in HBr, while Zn and Cu are not absorbed. Selectivity is also achieved in the elution step.

A new technique called "reactive ion exchange" can be used to collect trace elements from water.[8,9] It consists of incorporating in the ion-exchange column an ion that will react selectively with the ionic species to be collected and fixing it in the column. For example, chromate ions in water are collected on a short column of Dowex®-50 cation-exchange resin loaded with ferrous ions. Chromate is reduced to Cr (III), which is held on the resin along with any chromium originally present as Cr (III). Recovery of chromium is 90% from water containing 50 μg/l. Another example is the absorption of trace ferro- and ferricyanides by a Cu (II)-loaded resin.[10]

Where special selectivity is desired, special custom-made resins can be used that carry functional groups like dithiocarbamate[11] or thioguanidine.[3] Paper impregnated with thioguanidine resin has been used to take up traces of gold and mercury from water.[12,13] Resin-impregnated papers have frequently been used as filters to collect trace metals from water. Analysis is usually done *in situ* by X-ray fluorescence.[14]

B. Organic Ions and Molecular Species

Important organic pollutants that form negative ions are nitrilotriacetic acid (NTA) and phenols. They may be absorbed selectively by passing the contaminated water through an anion-exchange resin. NTA and citric acids, being relatively strong acids, are absorbed from tap water and sewage effluents buffered with formic acid to pH 2.5.[15] After absorption, they can be stripped with concentrated formic acid and determined by gas chromatography. Ultraviolet detection of NTA is also used.[16]

Phenols, being weak acids, form anions only at high pH. In one study,[17] the water samples were adjusted to pH 12, then passed through a column of macroporous anion-exchange resin in the hydroxyl form. After absorption, the phenols were stripped with acid, acetone, and water, then measured by gas chromatography. Initial concentrations of phenols in the water samples were about 20 μg/l.

It is not necessary for organic compounds to be ionic in order to be absorbed on ion-exchange resins. Resins have a polymeric "backbone" that acts as a solid organic solvent in its own right. Resins based on polystyrene have a strong attraction for uncharged aromatic molecules. The function of the ionic groups is to become hydrated and cause the resin to swell and become permeable; the nature of the fixed ions and counter-ions is secondary. Yet the counter-ion does have an effect. It has been found that calcium-loaded cation-exchange resins absorb aromatic hydrocarbons and chlorinated aromatic compounds more effectively than resins loaded with singly charged ions.[18] A macroporous cation-exchange resin in its calcium form absorbs aromatic hydrocarbons from water at parts-per-billion concentrations; they are stripped by methanol, measured by gas or liquid chromatography, and recovered with efficiencies of 80 to 100%. An advantage of this technique is that negatively charged organic ions, like those of humic acids, are repelled by the fixed ions of the resin and are not retained.

III. SEPARATION OF RECOVERED COMPONENTS AFTER PRECONCENTRATION

A. Inorganic Ions

Once trace metals have been concentrated into a small volume of solution, ion exchange can be used in well-known ways to separate the metal ions (as well as nonmetal anions). The separation can often be carried out on the same column that was used for preconcentration. For example, Co, U, Zn, and Cd were collected together on an anion exchange resin column from water made 0.1 *M* in potassium thiocyanate, then eluted separately in this order by decreasing concentrations of hydrochloric acid (cadmium was eluted with nitric acid).[19] This example is only illustrative; many others could be cited. Selectivity in absorption and elution is often enhanced by using mixed aqueous-nonaqueous solvents.[20]

Phosphate ions from polluted water were absorbed on an anion-exchange resin column and different anionic species were selectively eluted.[21] Anion-exchange chromatography is, in fact, the only good way to distinguish between the bewildering number of oxy-anions of phosphorus and anionic, partially esterified phosphates.

Selenium in natural and polluted waters is being recognized as a problem. Selenium may be determined by flameless atomic absorption; however, interelement effects are great, and most metal ions either enhance or suppress the atomic absorption signal. Solutions containing selenite ions, SeO_3^{2-}, preconcentrated by evaporation, were passed through a hydrogen-form cation-exchange resin column to remove all interfering cations.[21]

B. Organic Ions and Molecular Species

High-resolution liquid chromatography on anion-exchange resin columns has been developed to a fine art for the analysis of body fluids, primarily at Oak Ridge National Laboratory in Tennessee. It was only natural to apply the method to the analysis of polluted waters. The Oak Ridge group concentrated water samples up to 10,000-fold by vacuum evaporation and freeze-drying, then applied the concentrates to analytical columns 1.5 m long packed with 13-μm-size resin,[22] They were eluted by acetate buffers and monitored by a combination of ultraviolet absorption and fluorescence. Over 100 organic compounds were separated over a 40-hr flow period, and some 50 of them were identified. Some were acids, others neutral molecules, including sugars. Clearly this method is not suitable for routine use, but it is definitive. Organic constituents of natural and polluted waters are so complex, however, that no one method can give a complete analysis, and routine water quality control requires simple, rapid methods.

The analysis of trace organic constituents in water is a field that is expanding fast, and predictions are risky. It seems that ion-exchange methods will have a place, but a limited one. Non-ionic absorbents like the Amberlite® XAD resins (macroporous styrene-divinylbenzene copolymers) and surface-modified silica particles (C_{18} bonded packings) are easier to use than ion-exchange resins, though not as selective. It is hard to beat the efficiency of microparticulate C_{18} bonded packings for analysis by liquid column chromatography. Bonded packings with ion-exchanging functions may also be used.

REFERENCES

1. **Florence, T. M. and Batley, G. M.**, Removal of trace metals from sea water by a chelating resin, *Talanta,* 22, 201, 1975.
2. **Blount, C. W., Channell, R. E., and Leyden, D. E.**, An improved technique for the preparation of pellets for X-ray spectrographic analysis on ion-exchange resins, *Anal. Chim. Acta,* 56, 456, 1971.
3. **Blount, C. W., Leyden, D. E., Thomas, T. L., and Guill, S. M.**, Application of chelating resins to trace element analysis of geological samples using X-ray fluorescence, *Anal. Chem.,* 45, 1045, 1973.
4. **Murata, M. and Noguchi, M.**, An ion-exchange resin-epoxy resin pelletization method for sample preparation in X-ray fluorescence analysis, *Anal. Chim. Acta,* 71, 295, 1974.
5. **Lochmueller, C. H., Galbraith, J. W., and Walter, R. L.**, Trace metal analysis in water by proton-induced X-ray emission analysis of ion-exchange membranes, *Anal. Chem.,* 46, 440, 1974.
6. **Korkisch, J. and Sorio, A.**, Ion-exchange procedures for determining trace elements in natural waters. V. Lead, *Talanta,* 22, 273, 1975.
7. **Korkisch, J., Goedl, L., and Gross, H.**, Ion-exchange procedures for determining trace elements in natural waters. VI. Zinc. VII. Copper. VIII. Molybdenum, *Talanta,* 22, 281, 289, 669, 1975.
8. **Pankow, J. and Janauer, G. E.**, Analysis for chromium traces in natural waters, *Anal. Chim. Acta,* 69, 97, 1974.
9. **Janauer, G. E., Ramseyer, G. O., and Lin, J. W.**, Selective separations by reactive ion exchange: general considerations, *Anal. Chim. Acta,* 73, 311, 1974.
10. **Ramseyer, G. O. and Janauer, G. E.**, Selective separations by reactive ion exchange. II. Preconcentration and determination of complex iron cyanides in water, *Anal. Chim. Acta,* 77, 133, 1975.
11. **Dingman, J. F., Gloss, K. M., Milano, E. A., and Siggia, S.**, Concentration of heavy metals by complexation on dithiocarbamate resins, *Anal. Chem.,* 46, 774, 1974.
12. **Green, T. E., Law, S. L., and Campbell, W. J.**, Use of selective ion-exchange paper in X-ray spectrography and neutron activation, *Anal. Chem.,* 42, 1749, 1970.
13. **Law, S. L.**, Inorganic mercury collection by a selective chelating resin, *Science,* 174, 285, 1971.
14. **Tackett, S. L.**, Ion exchange filter paper holder for X-ray spectrometry, *Anal. Chem.,* 43, 972, 1971.
15. **Aue, W. A., Hastings, C. R., Gerhardt, K. O., Pierce, J. O., Hill, H. H., and Moseman, R. M.**, The determination of parts-per-billion levels of citric and nitriloacetic acids in tap water and sewage effluents, *J. Chromatogr.,* 72, 259, 1972.
16. **Longbottom, J. E.**, Determination of nitriltriacetic acid by high-speed ion-exchange chromatography, *Anal. Chem.,* 44, 418, 1972.
17. **Chriswell, C. D., Chang, R. C., and Fritz, J. S.**, Chromatographic determination of phenols in water, *Anal. Chem.,* 47, 1325, 1975.
18. **Ordemann, D. M. and Walton, H. F.**, Liquid chromatography of aromatic hydrocarbons on ion-exchange resins, *Anal. Chem.,* 48, 1728, 1976.
19. **Korkisch, J. and Goedl, L.**, Ion-exchange procedures for determining trace elements in natural waters. IV. Uranium, cobalt and cadmium, *Talanta,* 21, 1035, 1974.
20. **Korkisch, J. and Sorio, A.**, Determination of seven trace elements in natural waters after separation by solvent extraction and anion-exchange chromatography, *Anal. Chim. Acta,* 79, 207, 1975.
21. **Henn, E. L.**, Determination of selenium in water and industrial effluents by flameless atomic absorption, *Anal. Chem.,* 47, 428, 1975.
22. **Pitt, W. W., Jr., Jolley, R. L., and Scott, C. D.**, Determination of trace organic contaminants in natural waters by high-resolution liquid chromatography, *Environ. Sci. Tech.,* 8, 1068, 1975.

Uses of Polymers with Active Sites

Chapter 12

COAGULANTS AND FLOCCULANTS

L. B. Luttinger

TABLE OF CONTENTS

I. INTRODUCTION

While the terms "coagulation" and "flocculation" are often used interchangeably in the waste and water treatment fields, they are best considered as two distinct phenomena. In the field of chemistry, most authors follow the lead of LaMer,[1] who defines coagulation as destabilization produced by compression of the electric double layers surrounding small particles, and flocculation as destabilization produced by the formation of long-chain molecular bridges between particles. While coagulation and flocculation both result in the separation of a discontinuous phase, the properties of the resulting solids are different. Ideally, coagulation (brought about by the presence of a soluble salt) produces a coagulum composed of densely packed particles; filtration through a bed of such material is apt to be slow. On the other hand, flocculation, brought about, for example, by minute quantities of organic high molecular weight polymers, tends to produce a floc consisting of open-structured, loosely packed particles; filtration proceeds rapidly through a bed of such material. Many intermediate situations are known in which the phase separation cannot readily be defined as either flocculation or coagulation. However, in coagulation, proper discharge of the electric double layer around a particle is the central fact and is brought about by ions of a charge opposite to that of the net charge of the double layer, while in flocculation, the basic feature is the formation of long-chain bridges between particles. These bridges consist of polymer molecules which often carry the same charge as the particles being flocculated. The open, flocculated structure often produced is caused by (a) the presence of intervening chains and (b) mutual electrostatic repulsion of the particles due to the fact that their electric double layers need not be at least partly discharged by the polymers, in contradistinction to the situation obtained in coagulation.

We will follow the practice of considering as coagulants those materials which bring about coagulation as defined above and considering as flocculants those materials which bring about flocculation as defined above. Confusion in this area is abetted by factors such as the common industry tendency to refer to flocculants as "coagulant aids" and to use the terms coagulation and flocculation interchangeably.

While the use of coagulants is very old, going back at least 2000 years to when alum and lime were first used, the use of high-molecular-weight synthetic organic flocculants dates back only to about the mid-1950s when such materials first became generally available.

II. TYPES AND EXAMPLES OF COAGULANTS AND FLOCCULANTS

A. Coagulants

The simplest type consists of monovalent salts such as NaCl. The salts of divalent cations such as Ca^{++} are about 30 to 40 times as effective as these, on a same-weight basis, when used to coagulate negatively charged colloids (most colloids are negatively charged in water). The comparable salts of trivalent cations such as Al^{+++} and Fe^{+++} are about a thousand times as effective as NaCl on a weight basis, when used under conditions where salt hydrolysis does not occur. However, at normal pHs of usage, trivalent and some divalent metal salts are hydrolyzed and are even more effective, as discussed below.

Besides lime, alum, and iron salts, the only other materials in common use are activated silica and magnesium hydroxide, which like alum form polymers and/or flocs under the conditions of use. All these materials, except lime (Ca^{++}), possess some properties of flocculation as well as coagulation. Sulfuric acid can be used for certain coagulation applications such as emulsion breaking. In some cases, the acid acts to dissolve a "skin" on the surface of the emulsified particles. Table 1 lists some of the more common coagulants used in waste treatment.

B. Flocculants

All "simple" flocculants in current use are relatively high molecular weight organic compounds. Alum and iron salts are not included in this category since their behavior is more complex. A large proportion of these materials, regardless of other properties, are based on the polyacrylamide structure

TABLE 1

Coagulants Used in Waste Treatment

Material	Active structure	Remarks
Calcium ion (added as lime)	Ca^{++}	30 to 40 times as effective as NaCl on negative colloids on an equivalent weight basis; insensitive to pH
Sulfuric acid	H^+, $SO_4^=$	Useful for low pH applications, some emulsion breading
Alum	Alum "Polymers" such as $[Al_8(OH)_{20}]^{4+}$, etc.	Far more effective than Ca^{++} on negatively charged colloids, mainly useful in pH range 5 to 7; also has flocculation properties
Iron salts	Fe^{+++} "Polymers" such as $[Fe_2(OH)_2]^{4+}$, etc.	As above for alum; useful in pH range 4 to 11; also has flocculation properties
Activated silica	"Polymers" of unknown structure, probably based on −Si−O−Si−O−Si−O−	Probably has flocculation properties
Magnesium hydroxide	$Mg(OH)_2$ hydrous gel; behaves somewhat like alum	Useful in pH range above 11; probably has flocculation properties

$$-(CH_2-\underset{\substack{| \\ CONH_2}}{CH})_x-$$

where x is a large number. The predominant role of polyacrylamide derivatives among flocculants is based on (a) economic considerations, (b) on the fact that very high molecular weights are attainable, (c) that the basic structure may be easily altered, and (d) that the basic structure forms strong bonds to various solids. However, as discussed below, a number of other basic polymer structures are also in common use.

Flocculants may be conveniently separated into three types according to the overall charge on the molecule. These are anionic, cationic, and nonionic.

1. Anionics

Anionic flocculants represent the "purest" case of flocculation behavior unmixed with coagulation behavior as defined above. This is because, in most cases, the anionic polymer is used successfully to flocculate anionic (negatively charged) materials. Therefore, discharge of the double layer or charge neutralization can play no part here, and interparticle bridging by the long polymer chain is the only mechanism involved in destabilization. Among anionic flocculants, flocculation effectiveness is usually highly dependent on molecular weight, with the most effective flocculants (on a dosage weight basis) being those with the highest molecular weight. The latter may be as high as 20 to 30 million in some cases. Much lower molecular weight materials are often effective, however, depending on the application. The charge level of most of the anionic flocculants is usually quite low; most often, only a small percentage of ionizable groups such as carboxyls

$$\underset{\substack{| \\ COOH}}{-}$$

are present in the chain, and these are usually only partly ionized at the pH of use. However, a few anionic flocculants do possess a relatively high charge, due either to a preponderance of carboxylic or fully ionized sulfonic acid groups

$$\underset{\substack{| \\ SO_3^-}}{-}$$

Most anionic flocculants are based on the polyacrylamide structure, with a few percent carboxyl groups dispersed among the predominant amides.

2. Cationics

In this group of flocculants, coagulation and flocculation phenomena are often intermixed. This is because the cationic chains can enter and discharge the anionic double layer surrounding most colloid particles. Characteristically, polymers of rather low molecular weight (20,000 to 60,000) are often effective. A number of these materials are based on the polyethylene imine structure

$$-(CH_2-CH_2-NH-CH_2-CH_2-NH)_y-$$

where y may be the order of 1000. However, other cationics are based on a fundamental polyacrylamide structure containing interspersed cationic functionalities such as amines or quaternized amines

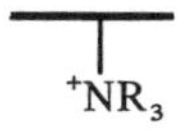

Some of these materials are of very high molecular weight, and apparently function predominantly via flocculation ("bridging") mechanisms.

3. Nonionics

The appelation "nonionics" with reference to flocculants is almost always a misnomer. Virtually all nonionic polymers carry a slight negative charge in water. Even when no ionizable groups are present in the polymer molecule (as in polyvinyl alcohol), sorption of a small number of negative ions from solution, conferring a negative charge on the polymer, will occur. Most of the important nonionic flocculants contain even more of a charge than this, since they are usually polyacrylamides containing a very small number of carboxyl groups originating from amide hydrolysis. The charge level is lower than that of anionic flocculants of the same type, but the difference is one of degree rather than of kind. Probably the closest things to true nonionics are those polymers of low gross charge (e.g., not proteins) in solution at pH values equivalent to their isoelectric points. At such points, positive and negative charges on the molecule are equal (the net charge is zero). For most organic polymers, the isoelectric point is at pH levels below the usual practical applications of these materials. However, this is not true for other materials such as alum and ferric alum. Practically speaking, nonionic polymeric flocculants can be considered as a gradation of anionics. This also corresponds to the manner in which they behave in practice. Table 2 lists some typical organic polymeric flocculants and their properties.

III. MECHANISM OF COAGULATION AND FLOCCULATION

A. Coagulation

The mechanism of coagulation of simple coagulants such as most monovalent and divalent salts acting on hydrophobic particles consists essentially of the partial discharge of the electrical double layer surrounding the colloid particle, by ions of opposite charge. As a consequence of this charge reduction, the interparticle repulsion potential is decreased. The particles can then approach one another more closely, leading to an increased number of collisions and therefore an increased coagulation rate.

B. Flocculation

The mechanisms of flocculation is generally much more complex than that of coagulation. At least two basic processes are involved in flocculation. These are attachment of the flocculant molecule to the particle and particle aggregation. Each of these, in turn, depends on several factors.

1. Mode of Attachment to Colloidal Particles

The most commonly used flocculants are polyacrylamide and its derivatives, and the most important mode of attachment of these and other water-soluble polymers to surfaces is probably hydrogen bonding.[2,3] For example, amide groups on such polymers can attach themselves via hydrogen bonding or bridging to oxygen atoms or hydroxy groups on surfaces such as that of silica

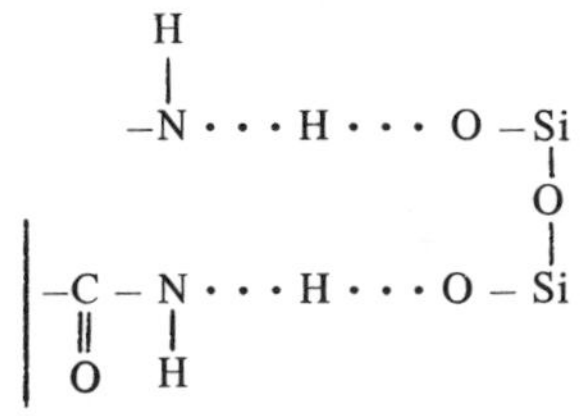

TABLE 2
Typical Organic Polymeric Flocculants

Material	Structure	Charge type	Remarks
Polyacrylamide	$-(CH_2-CH)_x-$ with $CONH_2$ on CH	Anionic (low charge type) and nonionic	Charge due largely to a small proportion of charged groups such as COO^-; charge depends on pH
Polyacrylamide	$-(CH_2-CH)_x-$ with $CONH_2$ on CH	Cationic (low charge type)	Charge due to presence of some cationic groups such as $-NH_2R^+$; charge depends on pH
Polyetheylene imine	$-NH-CH_2-CH_2-NH-$	Cationic (low charge type)	Charge due to H^+ $-NH-$ groups; charge depends on pH
Polystyrene sulfonic acid	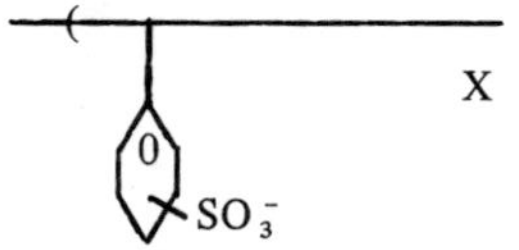	Anionic (high charge type)	Charge due to sulfonic acid groups; charge level not sensitive to pH
Quaternized ammonium	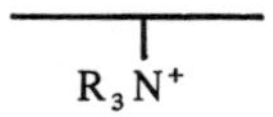	Cationic (high charge type)	Charge due to highly ionized R_4N^+ groups; charge level not sensitive to pH

Hydrogen bonding occurs between species containing O, N, or F atoms to which hydrogen atoms are attached. The bonding strength per bond of such moieties is generally low, being only a few percent at best of that of normal covalent bonds. However, in flocculation, thousands of such bonds may form between, for example, the amide groups of a polymer molecule and the enormous number of hydrogen bonding sites on even a very small colloidal particle. The overall bonding strength can therefore be enormous, in spite of the weakness of the individual bonds. Basically the same type of hydrogen bonding is involved in the great tensile strengths of certain fibers, such as those of cellulose, which are multilaterally bonded in the same manner.

Other attachment mechanisms are also possible. In electrostatic binding, a polymer attaches itself to a particle of opposite charge. Cationic polymers may bind to anionic surfaces in part by this means, i.e., the binding of polyethylene imine to negatively charged silica and clay surfaces. Also possible, but perhaps less frequently discussed, are specific chemical reactions between a functional group on a polymer and, for example, a divalent metal ion on or in a colloid surface. Carboxylate, sulfonic, or amine groups on a polymer may form complexes with suitably placed metal atoms on a surface.[4-8] In these examples, strong electrostatic or covalent binding may be involved.

2. Mechanism of Particle Aggregation

The crucial step in particle aggregation via flocculation consists of attachment (as in section 1 above) of two or more colloid particles to the same polymer chain. The particles thus attached are much more likely than unattached particles to overcome mutual electrostatic repulsion and to suffer further collisions. This leads to particle destabilization and to flocculation, i.e., the formation of comparatively loosely held aggregates or "flocs" (as opposed to coagula) containing large interstices, and therefore rapid settling, easy filtering and free draining.

Since attachment of a polymer chain to two or more particles plays so central a role here, those polymers that can form the longest "bridges" or "loops" should be the most effective, other factors being equal. This is observed within any one polymer class. The longest chain (highest molecular weight) polymers are generally the most effective materials, on a weight basis. This is particularly true for anionic and nonionic flocculants. Over the past few years, the molecular weights of the most important commercial flocculants have tended to increase, from about 10^6 in 1956 to about 2 to 3×10^7 15 years later. Still higher molecular weights would be used, except for manufacturing and dissolution problems. For cationic materials, where flocculation often involves electrostatic interaction as well as chain bridging, molecular weight tends to play a somewhat smaller, but often still important, role.

The extent of charge on the polymer also affects flocculation. For anionic polymers, there is a tendency toward diminished effectiveness, in the most straightforward applications, as the negative charge increases above a low level. This effect undoubtedly involves repulsion between highly charged negative polymer chains and the (usually) negatively charged particle surfaces generally found in aqueous systems. Apparently, a low level of charge on the polymer is desirable. This is possibly due to the fact that this negative charge tends, by intra-chain repulsion, to stretch out the polymer into a relatively more linear configuration, in which it is better able to collect particles. (Most polymers tend to adopt a random-coil configuration in solution, i.e., they are not highly extended in the absence of electrostatic charges along the chain.) However, when the charge on the chain reaches a certain level, repulsion between polymer and particle usually becomes important. It should be noted, however, that some very highly charged anionic polymers, e.g., polysulfonic acids, are good flocculants in many applications.

Cationic polymers usually do not function in quite the same manner in this regard, since the interaction is between a particle and a polymer of opposite charge; thus, charge repulsion between the two does not occur until high levels of particle coverage by polymer are attained. Generally, the higher the cationic charge, the more "stretched" the molecule (better particle collection) and the stronger the attachment between polymer and particle, at least up to the point of charge reversal of the particle due to polymer sorption. However, quantitative data are apparently not available in this area.

C. Effects of Polymer Fine Structure

Polymers of the same general type, i.e., polyacrylamides, which are very similar in molecular weight and charge level, often display important differences in flocculation power toward certain substrates. To be considered here is the fine structure of the polymer, i.e., extent of polymer chain "branching", presence of a small number of special groups, etc.

D. Hydrous Metal Oxides

As noted earlier, at pH levels (<3) below which hydrolysis does not occur, ferric and aluminum ions behave simply as trivalent coagulents. However, the important applications of these reagents occur at higher pH values, at which multicharged polymers of aluminum and ferric ions predominate.

These polymers, which are usually positively charged under the conditions of use, function in part by discharge of the double layer of negatively charged colloids. However, the polymer chains may also bridge to some extent between particles. This results in a flocculation-like effect. The formation of polymers and the occurrence of bridging are reflected in the fact that smaller dosages of the hydrolyzed polymers are required to effect colloid particle destabilization than are required in the pH range where aluminum and ferric ions exist as simple trivalent ions.

Actually, it is difficult, and probably not useful, to try to categorize these materials as coagulants and/or flocculants, since their behavior shares features of each and also possess certain features not shared with the simpler coagulants and flocculants, such as a tendency to form "sweep floc" and "pinpoint floc" and to collect dissolved as well as colloidal organic material, both hydrophilic and hydrophobic. Complex formation between the aluminum and iron ions and certain groups such as carboxylate and phenate must play a role in some of these reactions, as in the well-known aluminum/Aluminon "lake" reactions of analytical chemistry. Other hydrous metal salts, such as magnesium and titanium hydroxides, probably function somewhat similarly.

IV. APPLICATIONS IN WASTE TREATMENT

A. Coagulants

Coagulants, e.g., lime, alum, ferric salts, etc., are used in the treatment of a very wide variety of wastes. These include oily and greasy wastes, food, sewage, paint and pigment, metal, clay, textile as well as many other types. Lime is often employed for both coagulation and pH adjustment. A variety of coagulation methods is employed in these fields. Coagulation/flocculation is used in various settling (sedimentation) processes, particularly employing alum, but is also used in flotation, where a low-density waste is frequently alum-coagulated prior to aeration. Coagulation/flocculation also plays a role in waste treatment via filtration, i.e., in "in-line coagulation" via alum or iron salts. The role of coagulation/flocculation in the dewatering of sludges such as water plant sludges and other wastes which are thereby concentrated to the level of 5 to 50% solids is discussed in the following chapter by Schwoyer, Volume II, Chapter 13.

B. Flocculants

The employment of flocculants in waste treatment is as broad in its range as that of coagulants, and frequently coagulants and flocculants are employed together (although fed separately). In a number of cases, however, flocculants, particularly polymeric cationics, have been employed in place of positively charged coagulants such as alum. In such applications, the dosages of polymer required may be of the order of 1 to 5% of those required using alum or iron salts. However, since the cost of many cationic polymers is greater than that of the inorganic salts by roughly the same factor, utilization of these polymers in place of inorganic salts must be made on an individual basis. Two polymers (e.g., an anionic plus a cationic) are sometimes found to give superior results.

It is frequently found when using organic polymers that critical dosage levels exist, beyond which poorer and even negative results (colloid restabilization) are obtained. The correct dosage must be arrived at by experiment.

C. Examples

A few illustrative examples of the use of coagulants and flocculants on specific wastes are given below. It should be noted that dosages may vary considerably with the individual waste sample, e.g., of sewage, as well as with the individual polymer used.

1. Sewage (settling): A concentration of 0.15 ppm anionic polymer increased suspended solids removal to 78%, where "no chemical treatment" removal was 58%.
2. Paint and bonderizing waste (settling): 1 ppm organic polymer plus 170 ppm lime was claimed to give better performance than 34 ppm alum plus 200 ppm lime.
3. Oily refinery waste (flotation): A concentration of 7 ppm of a cationic organic polymer gave 60 to 95% removal of oil (down to ca. 10 ppm oil).
4. Paper mill waste (settling): 20 ppm cationic high polymer removed 60% more of the suspended solids, as compared with "no polymer."
5. Venturi scrubber clarification (steel) (settling): 2.5 ppm of an anionic polyacrylamide (low charge level) increased suspended solids removal to 85%; 2 ppm of the same polymer plus 6 ppm alum increased solids removal to 96%.

V. COMMERCIALLY AVAILABLE FLOCCULANTS

Table 3 lists some representative and commonly used polymeric organic flocculants. The list is necessarily abbreviated, as hundreds of different materials are currently available. Apparently, many small companies in this business merely repackage or resell a few standard products of the larger manufacturers, under various labels. On the other hand, some of the larger manufacturers may offer as many as 30 to 40 different flocculants.

VI. METHODS OF EVALUATION

The most common method of evaluating coagulants and flocculants for settling applications is by means of "jar tests." Typically, six beakers of the same waste sample are evaluated simultaneously. The idea is to straddle the optimum dosages of coagulant or flocculant in a series of tests. For example, if the approximate dosage required is expected to be between 20 and 200 ppm alum, 20, 40, 80, 150, and 200 ppm of alum

TABLE 3

Commercially Available Organic Polymeric Flocculants

Name	Structure	Charge type	Molecular weight	Remarks	Manufacturer
Alcolac® DV529	Polyethylene imine (PEI), somewhat more branched than DV530	Cationic	50,000 (?)	Strong chelate former with transition metals	Alcolac Inc., Baltimore, Md.
Alcolac DV530	Polyethylene imine	Cationic	50,000 (?)	Strong chelate former with transition metals	Alcolac
Betz® 1260	Polyacrylamide copolymer	Cationic	Very high (10,000,000?)		Betz Laboratories, Inc., Travose, Pa.
Cat-floc® B	Amine?	Cationic, moderate charge			Calgon Corporation, Pittsburgh, Pa.
Cat-floc T	Quaternary ammonium	Cationic, very high charge	500,000?		Calgon
CMC®-7MT	Carboxymethyl cellulose	Anionic	~1,000,000(?)	Rather stiff chain, metal complex former	Hercules Inc., Wilmington, Del.
Hercofloc® 814-2	Polyacrylamide (?)	Cationic	Above 3,000,000		Hercules Inc.
Hercofloc 828-1	Polyacrylamide (?)	Cationic	?		Hercules Inc.
Hercofloc 836-2	Polyacrylamide	Anionic	12,000,000		Hercules Inc.
Jaguar® Plus	Guar gum (related to starches) containing cationic sites	Cationic	250,000		Stein, Hall & Company, Inc., New York, N.Y.
Lytron® 820	Styrene-maleic acid copolymer	Anionic (high charge)	20,000	Somewhat hydrophobic, forms metal chelates	Monsanto Company, St. Louis, Mo.
Lytron 898	Vinylacetate-maleic acid copolymer	Anionic (high charge)	20,000 (?)	Somewhat less hydrophobic than above, forms metal chelates	Monsanto
Magnifloc® 521C	Polyacrylamide	Cationic (low charge)		Amine	American Cyanamid Company, Wayne, N.J.

TABLE 3 (continued)

Commercially Available Organic Polymeric Flocculants

Name	Structure	Charge type	Molecular weight	Remarks	Manufacturer
Magnifloc 575C	Polyacrylamide	Cationic (high charge)		Quarternary amine groups	Cyanamide
Magnifloc 820A	Polyacrylamide	Anionic (intermediate charge)	6,000,000		Cyanamid
Magnifloc 835A	Polyacrylamide	Anionic (high charge)	15,000,000		Cyanamid
Magnifloc 837A	Polyacrylamide	Anionic (low charge)	15,000,000		Cyanamid
Magnifloc 870A	Polyacrylamide	Anionic (very high charge)	200,000		Cyanamid
Magnifloc 905N	Polyacrylamide	Nonionic (actually slightly anionic, less so than 837A)	15,000,000		Cyanamid
PEI 7504-01	PEI, 50% quaternized	Strongly cationic	60,000 +	Highly charged over the entire pH range; little or no metal complex formation; possibly sterically hindered	The Dow Chemical Company, Midland, Mich.
PEI 7504-02	PEI, 75% quaternized	Strongly cationic	60,000 +	Has same characteristics as PEI 7504-01, but more so	Dow
Percol LT® 20	Polyacrylamide	Nonionic	—		Allied Colloids, Ridgewood, N.J.
Percol LT 22	Polyacrylamide	Cationic			Allied Colloids
Percol LT 25	Polyacrylamide	Anionic	—		Allied Colloids

TABLE 3 (continued)

Commercially Available Organic Polymeric Flocculants

Name	Structure	Charge type	Molecular weight	Remarks	Manufacturer
Primafloc® A-10	Copolymer of methacrylic acid and an acrylic ester in molar ratio of 2:1	Anionic	2,000,000—3,000,000	Supplied as an emulsion; forms complexes with metal ions including Ca^{+++}, Mg^{++}; most effective at alkaline pH	Rohm & Haas Company, Philadelphia, Pa.
Primafloc C-7	Polyamine · HSO_4	Cationic (fairly high charge)	1,000,000—2,000,000	Homopolymer, forms chelates with transition metals	Rohm & Haas
Purifloc® A-21	Polystyrene sulfonic acid	Anionic (very high charge)	—		Dow
Purifloc N-17	Polyacrylamide	Nonionic	Very high		Dow
Swift 5V	Protein Polypeptide	Amphoteric (anionic above pH ~5)	20,000	Animal glue (protein); metal complex former, somewhat hydrophobic	Swift and Company, Chicago, Ill.
Tydex® 12	Polyethylene imine	Cationic	60,000	Strong chelate former with transition metals	Dow
Wisprofloc® 20	Causticized potato starch, cross-linked with borate	Anionic	Very high	Contains 0.1% phosphorus in phosphate groups which bind to Ca^{++} (at ends of branched chains); web-like structure	Gamlen Chemical Company, South San Francisco, Ca.

may be added to successive beakers, under rapid stirring (ca. 90 rpm; paddle stirrer). After 1 to 5 minutes of rapid stirring, the stirrer speed is reduced to, e.g., 10 to 20 rpm for 10 to 60 min. Stirring is then stopped and the beaker contents evaluated for rapidity of settling, color, or turbidity removal, etc. Several such tests may be required to develop an optimum coagulant dose. When alum or iron salts are used, it is also usually desirable to determine an optimum pH. This will require pH control via alkali, lime or acid addition and a further series of jar tests to this end. While alum and iron are usually acidic, alum is often added as the alkaline aluminate. Finally, if the waste source is a large one, economics trade-offs between supernate (effluent) quality and costs will often be made.

Combined alum and polymer jar testing is often performed. The polymer is usually added to a fixed near optimum level of alum. Since polymers are far more expensive than alum, and overdosage with polymers can cause difficulties, the concentration ranges of the polymers must be examined more carefully in jar testing.

In some cases, jar tests will be applied to systems in which organic polymeric flocculants are the only reagents. Generally, these materials are less sensitive to pH variations than are alum and iron salts.

When coagulation is used in flotation applications, evaluation can be made by means of a pressurized "bomb" in which air release causes the coagulum to float, or by air passage through a glass frit into a liquid sample. The evaluation may also include a weighing (after drying) of the float or a measurement of the turbidity of the "undernate" effluent.

Coagulation/flocculation effectiveness in "in-line coagulation" can be evaluated on test columns via (a) effluent turbidity level, (b) gallons filtered to a certain turbidity breakthrough level or to a certain pressure drop across the filter, (c) filtration rate as in (b), etc. It should be noted that other information, such as backwashability of the filters, must also be assessed to fully evaluate a coagulant or flocculant in in-line coagulation.

A number of other evaluation methods cannot be reviewed here due to space limitations.

VII. RECENT ADVANCES

A. Coagulation

Probably the most important new coagulant is $Mg(OH)_2$, as developed by A. P. Black and associates.[9,10] Magnesium hydroxide is a highly hydrated, gelatinous material, possessing properties much like those of hydrolyzed alum and iron salts. Magnesium hydroxide is only precipitable at high pH (approximately 11). In practice, water containing magnesium carbonate is treated with lime until a mixed precipitate of magnesium hydroxide and calcium carbonate forms:

$$MgCO_3 + CaO + H_2O \rightarrow CaCO_3 + Mg(OH)_2 \qquad (1)$$

The dense calcium carbonate "weighs" the magnesium hydroxide to form a rapidly settling mixture. Postcarbonation of the sludge with carbon dioxide dissolves the magnesium. This may be recycled. The residual solid consists largely of calcium carbonate and is more easily dewatered than corresponding alum sludges. Lime and carbon dioxide may be recovered for reuse in large plants. While magnesium hydroxide has thus far been used principally for water plant treatment, it should also prove useful for various waste treatment problems such as organic color, turbidity, iron and manganese removal from and disinfection of industrial effluents.[10]

B. Flocculation

1. Several flocculants, of types which had previously proved difficult to dissolve, are now available in more convenient emulsified forms for easier handling.
2. At least two flocculants have been developed which are claimed to be useful in sea water or other high salt solutions.[11,12]
3. Hydrophobic and shear flocculation: Particles which have been hydrophobically coated, e.g., with soaps or with partially hydrophobic polymers (see Table 3) should be destabilized more easily on this score.[13]

C. Magnetically Abetted Separations

Magnetically abetted separations, i.e., filtration or coagulation on magnetically susceptible materials, have recently attracted considerable interest for waste treatment and other classes of separation.[14,15] Applicable reagents include alum, ferric salts, and polyelectrolytes. Reaction is made to take place around magnetically susceptible particles which are then removed from the liquid by a strong magnetic field. Economics are reported to be good for large scale (10 mgd) installations.

VIII. FUTURE PROSPECTS

During the last 6 or 7 years, there have been no real breakthroughs in this field, i.e., no new flocculants having decisively superior properties have become available. This is probably related to the fact that materials of higher molecular weight have not been forthcoming, i.e., the highest molecular weights of present (1979) flocculants (ca. 20 to 30×10^6) are about the same as those available 8 years ago. This is less true for cationics than for anionics and nonionics. The economics of processing higher molecular weight materials, plus the difficulties often involved in dissolution and mixing of the highest molecular weight types, are undoubtedly significant here.

Superior organic flocculants may be achievable by approaches other than molecular weight increase. These approaches could include (a) the development of polymers of greater hydrophobicity, to confer better dewatering properties, (b) the development of polymers with stronger binding groups, such as chelating groups where, for example, the colloid to be flocculated contains surface metal atoms, (c) the development of more branched or cross-linked polymers, wherein more than one polymer strand connects each of several colloid particles, thus making for overall stronger attachment and more sterically hindered conglomeration and therefore better dewatering, and (d) the development of more rigid, rodlike backbones, (e.g., cellulose and starch derivatives, of higher molecular weight). Such structures would tend to bridge over longer distances.

With regard to coagulants of the hydrous type such as alum, future prospects for new materials seem less bright. There are very few such materials which are effective, cheap, and nontoxic. However, a few materials investigated years ago such as the hydrous oxides of zirconium and titanium might be worth some investigation in this regard, in the light of modern knowledge of these materials and their pH-dependent behavior.

IX. CONCLUSIONS

The most significant development in this field during the last quarter-century has been the development of high molecular weight organic polymers. Used either alone, or in combinations of anionics and cationics, or in conjuction with alum and iron salts, they have tended to replace the older "coagulants" such as alum, although the latter still command a very large market. While comparison on a cost basis is not always favorable to the organics, the much smaller sludge volumes produced, the often superior effluent attained, and the frequent insensitivity to pH of organic polymers are real advantages which are expected to be reflected in ever increasing market penetration by these materials.

REFERENCES

1. **LaMer, V. K.**, Coagulation symposium introduction, *J. Coll. Sci.*, 19, 291, 1964.
2. **Michaels, A. S.**, Aggregation of suspensions by polyelectrolytes, *Ind. Eng. Chem.*, 46, 1485, 1954.
3. **Michaels, A. S. and Morelos, O.**, Polyelectrolyte adsorption by kaolinite, *Ind. Eng. Chem.*, 47, 1801, 1955.
4. **Luttinger, L. B.**, Studies on polyelectrolyte-metal complexes, Doctoral thesis, Polytechnic Institute of Brooklyn, 1954.
5. **Gregor, H. P., Luttinger, L. B., and Loebl, E. M.**, Metal-polyelectrolyte complexes. I. The polyacrylic acid-copper complex, *J. Phys. Chem.*, 59, 34, 1955.
6. **Gregor, H. P., Luttinger, L. B., and Loebl, W. M.**, Metal-polyelectrolyte complexes. II. Complexes of copper with cross-linked polyacrylic and polymethacrylic acids, *J. Phys. Chem.*, 59, 366, 1955.
7. **Anspach, W. M. and Marinsky, J. A.**, Complexing of Ni(II) and Co(II) by a polymethacrylic acid gel and its linear polyelectrolyte analog, *J. Phys. Chem.*, 79, 433, 1975.
8. **Summerauer, A., Sussman, D. L., and Stumm, W.**, The role of complex formation in the flocculation of negatively charged sols with anionic polyelectrolyte, *Kolloid Z. Z. Polym.*, 225, 147, 1968.
9. **Thompson, C. G., Yeary, C., Colyer, J., and Black, A. P.**, What's so great about $MgCO_3$?, *Water Wastes Eng.*, 41, August 1972.
10. **Thompson, C. G., Singley, J. E., and Black, A. P.**, Magnesium carbonate: a recycled coagulant, *J. Am. Water Works Assoc.*, 93, February 1972.
11. **Dow Chemical Company**, personal communication.
12. **American Cyanamid Company**, personal communication.
13. **Warren, L. E.**, Shear-flocculation of ultrafine scheelite in sodium oleate solutions, *J. Coll. Interface Sci.*, 50, 307, 1975.
14. **deLatour, C.**, Water purification by magnetic filtration, Paper No. 30A presented at the 2nd Natl. Conf. Complete Water Reuse, Chicago, Ill., May 8, 1975.
15. **deLatour, C.**, Process for removing dissolved phosphorus from water magnetically, U.S. Patent 3,983,033, September 28, 1976, assigned to Mass. Inst. Tech.

Chapter 13

SLUDGE DEWATERING WITH POLYELECTROLYTES

W. Schwoyer

TABLE OF CONTENTS

I. INTRODUCTION

During the past two decades, no single development in the area of sludge dewatering chemistry and equipment has had as profound an effect in a broad spectrum of applications as has the family of synthetic high molecular weight, water-soluble polymers classed as polyelectrolytes. They upgrade capacity of the process equipment, improve effluent clarity, and decrease the sludge volume when used alone or in combination with metallic salt coagulants in clarifiers. Polyelectrolytes improve the solids capture of a centrifuge and increase the capacity of drying beds, vacuum filters, and filter presses. They also increase the loading rates and run time duration in multimedia filters and make continuous mechanical gravity dewatering commercially economical.

II. ROLE OF POLYELECTROLYTES

Polyelectrolytes have two characteristics important for liquid-solids separation, whether in the initial stages such as clarification or in the later stages of sludge thickening and dewatering: electrical charge and molecular weight, as manifested in molecular length and structure. Usually as the electrical charge strength increases, the molecular weight decreases and vice versa.

The relative importance of the two properties varies not only with the type of separation being done, i.e., clarification vs. thickening vs. dewatering, but also with the surface charge and area of the particle to be treated. In the initial separation stages of clarification, whether it be by gravity settling or flotation where there are fine, sometimes colloidal size particles kept in suspension by repulsion of like charged surfaces, a highly charged coagulant capable of neutralizing the surface charge is required. Once the surface charge on the particles is neutralized, the particles must be encouraged to collide, forming larger particles which can then be further concentrated and dewatered.

After the surface charge has been neutralized, a high-molecular-weight (long chain molecule) polyelectrolyte is required for agglomeration of the colloid particles. The high molecular weight of polyelectrolytes, therefore, plays the predominant role in most dewatering applications.

As a rule of thumb, cationic polyelectrolytes have a higher charge and lower molecular weight than the nonionic or anionic species. The combination of high electrical charge and low molecular weight makes it important to determine the proper amount of polyelectrolyte to be fed in reaching the isoelectric point. An overdose can reverse the surface charge on the particles to the degree where they will once again disperse and become suspended due to the new like surface charge. This can happen either in the clarification, thickening, or dewatering phases of liquid-solids separation.

In practice, redispersion seldom is encountered when an overdose of an anionic polyelectrolyte occurs. This is because the anionic polyelectrolytes, which are usually derivatives of polyacrylamide, have a long chain length and a low charge. In these instances, the "bridging effect" encountered in the entanglement of the particles with the long chain polyelectrolyte molecules is stronger than the repulsion of like surface charges on the agglomerated polyelectrolyte/sludge coagulum.

In dewatering operations, the molecular weight of the polyelectrolyte is a more important consideration than encountered in the earlier liquid-solids separation steps of clarification and thickening.

III. SLUDGE CHARACTERISTICS

Very seldom does a sludge contain a single waste component by the time it is ready to be dewatered. As was mentioned earlier, the coagulants and coagulant aids are added in the clarification and thickening steps in order to increase the capacity and improve the effluent quality of the process equipment. Therefore, the sludges received for dewatering are usually a combination of the waste particles, coagulant, and coagulant aids. Because of economic considerations, the amount of coagulant aids added to the waste stream is usually the bare minimum required to render the waste particle manageable in the particular liquid-solids separation device chosen. As the solids concentration increases during the various mechanical separation stages, i.e., clarification to thickening to dewatering, so does the degree of physical handling. Usually the greater amount of physical handling encountered, the greater amount of coagulant aid, usually polyelectrolytes, required. The chemical conditioning requirements for a given waste in the clarification step will not impart sufficient integrity to the coagulum to withstand the rigors of mechanical dewatering. Even in the case of drying beds, a higher dosage of polyelectrolyte added to the waste stream than is required for settling in the clarifier will cause more rapid drainage on the sand bed.[1]

As soon as a second component is added to a pure waste, the dewatering characteristics for that waste stream will change. Therefore, as more solids are added, the characteristics of the waste changes. The handling characteristics of the coagulum can be considered a weighted average of the handling characteristics of the component parts.

The importance of this consideration is well demonstrated in the case of a combined ferric hydroxide/oily waste sludge. Due to its high positive surface charge, a relatively pure ferric hydroxide will require ca. 7 to 10 lb of an anionic polyacrylamide-based polyelectrolyte per ton of dry solids dewatered. When combined with the oily waste, the polyelectrolyte requirements can drop to as low as 2 to 4 lb/ton of dry solids.

Wastes consisting of a single, high surface charge component usually require a large amount of polyelectrolyte and are expensive to treat.

Concentration of the solids to be treated also has a marked effect on the polyelectrolyte dosage. A 9 wt% calcium carbonate sludge will require 0.5 to 1 lb polyelectrolyte per ton of dry solids, whereas a 3 wt% sludge of the same material will require 3 to 6 lb/ton of dry solids.

IV. CONDITIONING REQUIREMENTS

One of the most striking differences between the conditioning requirements for the lower solids level separation steps such as clarification and the higher solids loading encountered in dewatering is the amount of coagulants or coagulant aids used, whether it be polyelectrolytes, metallic salts, or inert fillers. These are used primarily for two reasons: the greater degree of mechanical handling required and the larger quantity of solids present in the dewatering step. Solids levels of 10 to 1000 mg/ℓ are considered for clarification, whereas those levels encountered in dewatering are at least an order of magnitude greater. The type of dewatering equipment to be used directly determines the type and quantity of coagulant aids to be added. Most types of dewatering equipment only use volumes to be processed. It is necessary that the flocculated agglomerate to be processed in the particular device have the certain consistency that will best operate in that equipment. The chemical conditioning required for the particular type of system will vary from device to device. It is necessary that this be obtained in order to have efficient liquid-solids separation.

The choice of conditioning for any specific device depends on what conditioning was done in the prior separation step. Here again, the amount of conditioning encountered will vary depending on the particular clarification or thickening device used. Solids in a slurry will tend to either sink or float once they are taken out of suspension. Clarification devices, therefore, can operate either by allowing the particles to settle or float. Usually, any given waste can either be settled or floated, depending on the type of conditioning system with which it is treated. Therefore, if a particle is treated in the clarifier using the natural tendency of the particle to either settle or rise, less conditioning is required than in the situation where a particle that tends to settle is floated or a particle that tends to float is settled during clarification. Here again, the coagulum that ultimately reaches the dewatering step will have characteristics of a combination of its various components.

Another consideration which bears on the conditioning requirements of the agglomerate floc particle delivered to the dewatering device are the ultimate requirements for the dewatering step. What is wanted? What must be the effluent suspended solids level? Is unit capacity of prime importance? Are cake solids more important than the first two characteristics? Usually excellence costs money. The clearer the effluent, the higher the unit capacity; the higher the discharge cake solids, the more conditioning required — hence, a higher cost.

The degree of treatment, and thus the cost involved, are usually dictated by the ultimate disposal method for the various components of the waste stream. If the sludge is going to landfill, a lower solids level may be required than if the material is to be incinerated. The most economical system of dewatering will be one in which the equipment and conditioning systems chosen will be those which deliver the specific degree of separation at the lowest cost.

At times, a single piece of equipment might have a wide enough operating latitude to meet different criteria. If not, a system should be chosen in which only the necessary amount of processing is done most effectively. A good example of this is a gravity/press dewatering system in which the cake discharged from the gravity unit is acceptable for landfill in most instances. When it is necessary to obtain a higher solids loading, the cake can then be further dewatered, usually to twice the solids content, in the press. With most sludges, there is no further conditioning required when the sludge is run through the press. For certain applications such as oily wastes and latex, a higher amount of conditioning is required when running the press than is necessary for adequate processing through the gravity unit.

Many of the colloids present in waste streams, i.e., oil, latex, some activated sludges and even metal hydroxides to an extent, are amorphous and do not contain the proper degree of integrity to withstand the mechanical handling encountered in the dewatering equipment, even though they may already contain a certain amount of coagulant aids added earlier in the initial clarification step. These wastes are frequently characterized by low ash and/or fiber content. When conditioning for dewatering, it is necessary to add materials which will decrease the tendency for the sludge particles to deform and become mobile under pressure. Here, again, the amount of coagulant aid required will be dependent upon the sludge particle itself, as well as the pressure to which it will be subjected in the dewatering process. The most commonly used coagulant aids are various clays, fibrous materials such as wood fibers, highly insoluble inorganic crystalline materials, and metal hydroxides.

Metal hydroxides are probably the most commonly used coagulants. They contain a high surface charge and are, therefore, excellent collectors of fine colloidal materials such as oil, latex, silt, and fine organic particles. They are also the least expensive of any of the charged materials used.[2] They are much more rigid than colloids such as oil and latex, but considerably more mobile than crystalline solids such as clay, precipitated inorganic salts, and paper fiber. One of the advantages of using the common metallic coagulant aids such as ferric chloride, ferric sulfate, or aluminum sulfate is that the addition of these materials to the waste results in a pH depression. In order to form the hydroxide, the pH must be raised to the 4 to 8 range. Best clarification is considered to take place at the pH range of 5 to 7. Practical experience with ferric and aluminum hydroxides indicates that the best dewatering takes place when the hydroxides are at the pH range of 6 to 7.

Neutralization of the sludge to which the acidic ferric or aluminum sulfates have been added is generally accomplished by the addition of either sodium hydroxide, calcium hydroxide (lime), or sodium carbonate. The addition of lime is desirable for dewatering because the insoluble calcium sulfate crystal is formed *in situ* within the hydroxide precipitated hydroxide/colloid waste coagulum, increasing the integrity of that combined particle with regard to applied pressure.

While most colloidal particles contain some degree of a surface charge, very few are at their isoelectric point. This is also true of the coagulant aids added in the clarification stage, whether it be clay used for weighting the particle or the metallic hydroxide and/or polyelectrolyte applied to form the coagulum and increase its size, due to surface charge neutralization and particle bridging.

In dewatering, the same chemicals (in greater amounts) used in the clarification step are frequently employed when obtaining the particle integrity required for a specific type of dewatering apparatus. Metal hydroxides, clay, fibers, and inorganic crystalline solids are used to increase the particle strength. The polyelectrolyte is used to flocculate the coagulum masses into a larger particle, usually ranging from 1/16 to 3/8 in. in mean diameter, squeezing out the interstitial water.

In the flocculation step, the bridging mechanism afforded by the polyelectrolyte's high molecular weight is usually of prime importance. As mentioned earlier, once the colloidal particles have been encapsulated into the larger coagulum in the clarification step, the effect and importance of the polyelectrolyte's charge diminishes. The degree to which it contributes to the effectiveness of the polyelectrolyte as a flocculant is dependent upon the net surface charges of the coagulated particles on which it operates. Net surface charge of the coagulated particle still dictates the choice of the polyelectrolyte, i.e., anionic, cationic, or nonionic. However, certain sludges are reported to have been flocculated equally well with anionic, cationic, or nonionic polyelectrolytes of the same molecular chain structure and weight.[3]

Nonionic high-molecular-weight polyelectrolytes are usually considered for use with coagulated particles whose surface charges are at their isoelectric point, or net zero charge. Frequently, the flocculum formed by the use of a nonionic polyelectrolyte is applied in the dewatering operation. In these instances, a dual polyelectrolyte system is employed: one charged polyelectrolyte followed by a second, oppositely charged. There are no hard fast rules dictating which charge polyelectrolyte is first added to the coagulated sludge. In most instances, the coagulated sludge particles maintain some slight surface charge. Usually, a polyelectrolyte having a charge opposite the net surface charge is applied. Enough of the first polyelectrolyte is added to impart a net surface charge sufficiently strong to attract the second, oppositely charged polyelectrolyte.

At the present time, the proper choice of polyelectrolyte is still most easily determined in a jar test. Zeta potential or streaming current detectors, while giving a good indication of net surface charge, by themselves tell little if anything about the strength of the flocculated mass, a property of utmost importance in dewatering.

Overdosing with polyelectrolyte can, at times, be as detrimental as insufficient addition. It was mentioned earlier that an excess of highly charged, low-molecular-weight polyelectrolytes, normally cationics, can cause redispersion of the coagulated particles due to repulsion of like surface charges, decreasing filter capacity, and increasing filtrate-suspended solids levels. Decreased filter capacity is also encountered with an excess of high-molecular-weight, low-charge polyelectrolytes, but not as a result of particle dispersion. In this case, the polyelectrolyte molecule is so long that it blinds the pores of the filter media, preventing passage of the loose water.

V. MIXING

When considering the use of coagulants and coagulant aids, a great deal of attention is paid to the proper selection of a specific reagent with regard to its surface charge and molecular weight. This, of course, is important, but almost equally important and usually neglected is the proper mixing of the flocculating material with the coagulated sludge. Both the coagulation of the initial colloidal material and subsequent flocculation of the coagulated particles are surface phenomena. It is, therefore, important that the reacting materials become thoroughly mixed if they are to form a particle having the greatest possible integrity. There are several factors which have a direct bearing on this situation:

1. The high-molecular-weight polyelectrolytes usually mixed at concentrations ranging from 0.25 to 2.5 wt% solids should be mobile enough to thoroughly mix with the sludge solids. This requires dilution of the polyelectrolyte to a concentration of 0.005 wt% for the very high molecular weight.
2. The solids in the influent sludge slurry must also be dilute enough to thoroughly mix with the polyelectrolyte.
3. The agitation applied must be sufficient to keep the particles suspended and cause contact with the high-molecular-weight polyelectrolyte, while not being so vigorous as to shear the flocculated particle.

Only recently have mixing studies specifically dealing with mixer design for the particular areas of coagulation and flocculation been carried out and reported by Hahn and Klute.[4]

VI. CONCLUSIONS

Polymeric entities with cationic or anionic groups as well as those with no apparent ionic groups, but capable of bridging, have become important aids in the volume reduction and dewatering of various types of sludges. The capacity of these polyelectrolytes for water reduction depends on the type of sludge, the devices used, and to some extent on the solids content of the mixture being treated.

The use of flocculant aids began about 30 years ago when certain starches aided in settling and filtering phosphate rock and when it was observed that clays, effective as aids in coagulation, had the properties of ion exchange. Today, many major chemical companies are manufacturing various synthetic polymeric materials for dewatering sludges. Both polymer types having varied molecular weights and active sites are available.

Solubility, especially of those polyelectrolytes with high molecular weight, is a problem to the user. Therefore, the trend is toward liquid polyelectrolytes, which in some cases require emulsification which can be demulsified with an emulsion prior to use.

Cost is another factor. Plants generating waste weigh the cost of hauling waste away against the cost of the polymers used to treat the waste. Lower cost products are being sought. These may take the form of synthetic polymers having a greater concentration of active sites per unit of polymer or modified natural products.

REFERENCES

1. **Nebiker, J. H. and Adrian, D. D.,** Cost evaluation of sand bed dewatering and drying of wastewater sludges, Filtr. Sep., p. 245, May/June 1969.
2. **Roberts, K. and Olsson, O.,** The Role of Colloidal Particles in Dewatering of Activated Sludge with Aluminum Hydroxide and Cationic Polyelectrolyte, presented at Annu. Meet. American Chemical Society, Chicago, Ill., August 24 to 29, 1975.
3. **Schwoyer, W. L.,** Dewatering of Lime Softening Sludges, Trans. 22nd Annual Conf. Sanitary Engineering, University of Kansas, Lawrence, 53, 53, 1972.
4. **Hahn, H. H. and Klute, R.,** Effect of Turbulent Flow Conditions Upon the Process of Coagulation, presented at Annual Meet. American Chemical Society, Chicago, Ill., August 24 to 29, 1975.

Chapter 14

POLYELECTROLYTE DISPERSANTS

J. F. Wilkes

TABLE OF CONTENTS

I. INTRODUCTION

Polyelectrolytes and other polymers with active sites find numerous applications in water treatment and pollution control, including the related aspects of water conservation, water recovery, and water conditioning for subsurface injection, such as oil field well drilling and water-flooding processes. Polyelectrolytes and polymers, both of synthetic and natural origin, are also valuable in specialized applications such as flow improvement and synergists for corrosion inhibition.

The use of both natural and modified polymers in water treatment has been in widespread use for many years. More recently, however, concentrated attention has been given to the evaluation of modified natural polymers and synthetic polyelectrolytes in the broad areas of flocculation, dispersion, and scale control, as related to systems using or recovering water or in which water is treated for pollution control.

II. APPLICATIONS

A. Steam Generation

In the earliest years of the Industrial Revolution, it was discovered that scale deposition and internal corrosion could be minimized in boilers by adding natural polymers such as starch (a polysaccharide with hydroxyl functional groups), natural tannins (polyhydroxyphenols having pyrogallol or catechol structures), lignin extracts, and numerous seaweed derivatives (mannuronates, alginates, agar, and other water-soluble gums.) Carboxymethylcellulose (a polysaccharride modified to introduce carboxyl functional groups) and lignin sulfonate derivatives, similarly rich in hydroxyl and carboxyl functional groups, are typical of newer modified natural polymers having broad application as dispersants for boiler feed water treatment and pollution control.[1-4]

In most internal treatment processes of boiler feed waters, inorganic agents are used to precipitate scale-forming solids in most desirable forms within the boiler. Organic polymers are added to prevent adherance of precipitates to heating surfaces and to complete the formation of fluid sludges, easily removed during blowdown. These organic substances are believed to function in preventing scale formation by several mechanisms. Their action may be dispersion, coagulation with adsorption of inorganic precipitates, coating action (which minimizes coherence and adherence to surfaces), plus distortion of crystal formation. Of major importance is the latter, which involves the ability of polymer functional groups to enter into chemical reactions with scale-forming minerals, causing incorporation of large, irregularly shaped bodies in the crystal lattices. The resulting distortion of the crystal lattices prevents deposition of dense, uniformly structured crystalline substances on the heat-transfer surface. In addition to the reduced adherence tendency of these irregularly shaped crystals, they probably develop internal stresses which cause the crystals to fracture and partially release from heated surfaces.[5,6]

Nonuniformity in the quality and chemical characteristics of tannins and other natural "polymers" and the operating problems encountered in practice (such as charring on overheated boiler metal) soon indicated the need for more stable, specifically structured polymers which could be tailored to meet industrial needs, incorporate desired functional groups, and control molecular weight. While some improvement was made by chemical modification of natural polymers, major progress was not achieved until the introduction of truly synthetic polymers and copolymers (polyacrylates, polymethacrylates, acrylamide-acrylic acid, and styrene maleic anhydride copolymers) beginning about 1950. The acrylic acid and methacrylic acid polymers are effective in very low dosages, ranging from 0.3 to 5 mg/ℓ in feed water, with optimum dosages of about 0.5 to 1.0 mg/ℓ[7-9] (For natural polymers, typical use dosages were 7 to 20 mg/ℓ in feed water.) The synthetics also could be used effectively with antifoam agents to maintain steam purity, conserve heat and water and reduce pollution potential by minimizing blowdown discharges. Antifoam activity and boiler life both were improved by the synergistic action of natural/synthetic polymer sludge conditioners.[10] Typical structures of natural, modified natural, and synthetic polymers being used in steam generation and for other water treatment purposes as shown by Denman et al.[2,3] and Sexsmith et al.[5] are illustrated in Figure 1.

B. Dispersants and Antifoulants in Cooling Systems

In view of the successful use of synthetic polymers in boilers, their adoption in cooling systems to replace formerly used natural organics was a logical step. Because of increasing industrial demand for energy and the large quantities of heat rejected, coupled with water shortages and a growing awareness of environmental problems, formerly used "once through" cooling systems were extensively replaced by high-duty open recirculating systems, including cooling towers and other evaporative systems. It was immediately noted that open cooling towers and associated cooling circuits were subject to serious problems of fouling, microbiological growth, and silt accumulation, traceable to the scrubbing action of cooling water on the vast quantities of air passing through the towers. In addition to airborne dust and silt and suspended solids entering

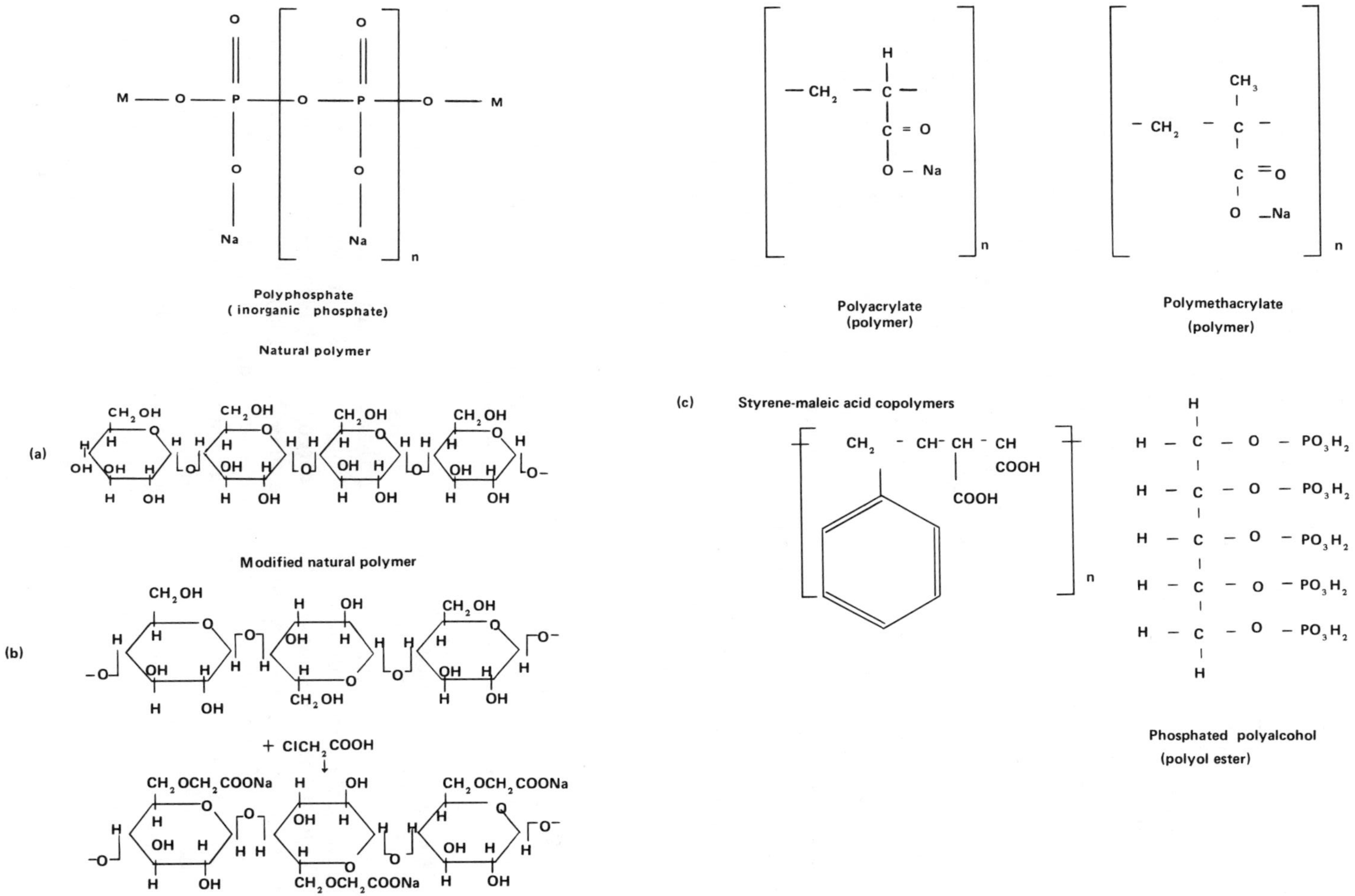

FIGURE 1. Polymeric structures.

with the makeup water, corrosion products and chemical reaction residues generated within the systems added to the load of suspended matter, as did dead microorganisms and their decomposition products. All these contaminants combined to form fouling layers which interfered with heat transfer and created local sites for intensified corrosion. To remove and disperse these fouling layers, maintain cooling system efficiency, and permit corrosion inhibitors to function requires carefully selected dispersants/antifoulants, in addition to corrosion inhibitors, scale control additives, and microbicides. These four additive groups work together synergistically; almost every cooling system will require their combined use.

It is essential to examine the causes of fouling and choose appropriate polymers and functionalities for specific problems encountered. Dispersion properties are important for control of corrosion products and the removal of dead biofoulants; flocculation and charge neutralization are desirable in silt suspensions to permit removal from the system by bleedoff. Cooling system scale control by polymers is similar to scale control in steam generators. Molecular weights in the general range of 10,000 and below appear most effective in cooling water scale control, as was also observed in steam generating systems. Higher molecular weight polymers may do a better job in silt suspension, suggesting the desirability of considering more than one polymer type and functionality where both types of problems exist. The maintenance of clean surfaces is vital to proper function of corrosion inhibitors and microbicides, and synthetic polymers play a major role in preserving "clean system" status. Obviously, polymer selection must consider the potential incompatibility with other dispersants and microbicides used in the system.[11] Recent publications have confirmed the previously reported utility of low molecular weight polyacrylates (such as CYANAMER® P-70) as scale control agents, when used at low dosages (5 to 15 mg/ℓ) in highly concentrated evaporative cooling systems, to promote pollution abatement and maximum reuse of industrial process water.[12] Low-molecular-weight polymethacrylate polymers, styrene maleic acid copolymers, and polyol esters all have utility in scale control cooling water treatment.

C. Mud Dispersion and Coalescence

For years, heavy industrial plants, such as steel mills, used available surface water supplies for cooling blast furnace jackets, open hearth furnace doors, and critical circuits in other production equipment. In the process, heavy deposits of mud, silt, and airborne process dusts often accumulated and caused overheating failures. Recognition of the utility of high molecular weight polyacrylamides and other water soluble polymers in lifting and dispersing such mud and silt accumulations can be traced to means developed for the correction of a similar problem of mud/silt accumulations in ballast tanks of Great Lakes ore carriers. All such cargo vessels have large ballast tanks which are used to control buoyancy when they are unloaded. Mud and silt entering the ballast tanks with water gradually accumulate in clay-like caked layers which remain behind when ballast tanks are emptied. Mud accumulations of as much as 1500 tons have been observed; they reduce cargo capacity of the vessel and so must be removed.

In the past, it was necessary to cut large openings into ballast spaces and manually shovel out or "hydraulically mine" and pump out the mud, both being unpleasant, laborious, and costly operations. W. E. Zimmie, a marine architect, conceived the novel approach of adding a polyelectrolyte which would penetrate, disperse, and suspend caked mud in fluffy, flocculant particles; being less viscous, these could be pumped out with ballast water. Working with F. W. Bloecher of American Cyanamid, Zimmie conducted extensive field research to evaluate and select suitable water soluble polyelectrolytes for mud removal use. A U.S. Patent covering the process was granted in 1963[13] The preferred flocculant stipulated in the patent claims was a polyacrylamide having a molecular weight range of 100,000 to 15 million, with an optimum molecular weight range of 4 to 6 million. (Many other polyelectrolytes have a structure similar to the =C=C′−R structure described in the patent specification.) Dosages stipulated for initial cleaning of mud-laden ballast tanks ranged from 0.01 to 5.0 lb polymer per ton of silt (dry basis), with a ratio of about 0.2 lb ton of silt being preferred. It was recognized that complete removal of very heavy deposits might require from three to 100 successive applications, but at substantially reduced dosages. Manual agitation or hydraulic stirring by hose streams and hydraulic jets were said to be helpful in dealing with heavy accumulations. A water depth of 6 to 36 in. above the mud layer was reported to produce an effective distribution of polyelectrolyte, with the 6-in. depth being preferred with normal vessel movement.

Once ballast tanks have been effectively cleared of mud deposits, further accumulations may be prevented by adding a flocculating agent to ballast water as it is pumped into the tanks. New silt will be suspended, and flocculated silt carried away when ballast water is discharged. Dosages will vary with the capacity of ballast tanks, but maintenance dosages of 1 to 20 lb per ballasting are reported effective, with a preferred addition about 10 lb per ballasting.

This same technique was then applied to remove silt deposits from inner surfaces of cooling jackets of large blast furnaces and from the doors of open hearth furnaces operated by the steel companies who owned the ore carriers. Here, the flocculant was added at average dosages of 1 mg/ℓ to once-through cooling water; the broad dosage range was 0.01 to 100 mg/ℓ. The open market price of the preferred polyacrylamide was ca. $1.75/lb, but applicable patent royalties and service fees covering the cleaning processes generally raised the chemical costs to about $6.00/lb for those operating under the Zimmie patent. It was later found that other acrylic-based polymers falling outside of the scope of the Zimmie patent claims also had utility for these and similar problems.

D. Mud Stabilization

Another interesting application for polymers with active sites is the dispersion and stabilization of mud for specialized services, such as the preparation of oil field well drilling muds. Sodium polyacrylates, such as Acrysol® GS (Rohm & Haas), Goodrite® K-708 (Goodrich), and similar commercially available products, with a molecular weight of about 100,000 have demonstrated their value as thickeners for synthetic latices and also for stabilization, fluidity, and viscosity control in drilling muds. Clay suspensions are pumped downhole continuously during drilling operations to seal formations along the bore, to cool drilling bits, and to remove rock chips from the drilling face. It is essential that the drilling muds flow freely, remain in suspension (instead of settling out in heavy, clay-like deposits), and retain sufficient density to suspend and remove sand and rock chips. In this empirical art, polyacrylates, low molecular weight acrylamides, and other acrylic copolymers, as well as modified lignin derivative polymers, are used extensively. For drilling mud applications, polymer dosages are measured in pounds per ton (dry)of drilling mud clays.

E. Subsurface Injection

Brines released during crude oil production represent a major pollution potential and create serious disposal problems. Ideally, it would be desirable to return the separated brine to the oil sands from which it was produced. This can be accomplished by the recovery of brine and extensive pretreatment to permit reinjection in water flood wells spaced around producing oil wells. Pretreatment of water or brine for injection is a complex process, involving clarification (coagulation and filtration); deaeration to minimize corrosion of injection lines; stabilization to prevent precipitation reactions with dissolved minerals in connate water; and possibly microbicide treatment (or chlorination) to prevent anaerobic bacterial growth which causes corrosion of lines and plugging of formations. Polymers have utility in the stabilization of injection water, but product selection and dosages must be based on carefully controlled field laboratory tests.

Subsurface injection of unwanted waste solutions has merit as a disposal method to prevent water pollution problems. However, before this can be attempted, clearance must be obtained from local, state, and Federal regulatory authorities, who generally are opposed to such disposal methods. The proposed receiving stratum must be shown to be tightly sealed above and below by nonporous rock layers to prevent contamination of potentially useful aquifers and must possess adequate porosity to receive the waste solutions. The receiving stratum and dissolved salts in its connate water must be of a mineral nature which will not react adversely with the waste solutions. Distances from nearest major users of deep well water supplies who might conceivably be affected also must be plotted accurately, since separations of 20 to 50 miles between disposal site and points of water withdrawal may be needed. Even when all these potential objections can be resolved, the fluid to be injected must be treated and clarified, when necessary, to prevent clogging the receiving rock formation. Polymers are applicable for treatment of disposal solutions, both for clarification and also for stabilization to prevent precipitation reactions with water (or brine) being displaced and with mineral constituents of the rock.

F. Steam Flooding

Steam flooding or stream injection for increased oil recovery is being practiced in many oil fields, using unique once-through coil-type boilers discharging high pressure wet steam (about 70% quality) directly to injection wells. In many cases, the only boiler makeup available is produced brine or water from poor-quality sources. Usually the feed water will be softened with multiple ion exchange columns or treated by special ion exchange processes to reduce hardness to the minimum. Oxygen scavengers, chelants, and polymers all may be used to prevent discharge of suspended solids downhole with the steam/water mixture delivered from the steam generator. Low molecular weight polymers (polyacrylate/methacrylate) similar to those used in conventional boiler operations are applicable here; slightly higher dosages may be appropriate.

G. Improved Corrosion Inhibition

It was noted previously that clean surfaces in cooling system heat exchangers respond most effectively to corrosion inhibition, allowing operation with minimum inhibitor residuals, substantially lowering pollution potential from cooling tower bleed-off water which must be discharged. All cooling water polymers which effectively control scale deposition and fouling or which lift and disperse mud and silt accumulations should lead to improved corrosion inhibition. This is a major application for active site polymers in cooling water treatment. Among those which have demonstrated value are polyacrylates, polymethacrylates, polyol esters, and hydrolyzed polyacrylamides of low molecular weights.

The application of polyacrylamides and related acrylic polymers for mud and silt dispersion also has been demonstrated to aid in the removal of corrosion product tubercles and inhibiting further tuberculation of ferrous surfaces.[14] Polymers and copolymers in the broad molecular weight range of 25,000 to 25 million have utility, with polyacrylamides in the molecular weight range of 4 million to 5 million appearing to have optimum activity. Use dosages range from 0.1 to 100 mg/ℓ (by weight), depending on the molecular weight of the selected polymer. A related patent[15] proposes a combination of two polymers, such as a polyacrylamide of fairly high molecular weight with a lower molecular weight polyacrylate or phosphonate, plus a water-soluble silicate or chromate compound. Hwa et al.[16] used a synergistic mixture of sodium polymethacrylate (9000 mol wt) and hydroxyethylidene diphosphonic acid to inhibit deposits of scale, sludge, and foulants in aqueous systems such as boilers, evaporators, and water cooling equipment, at substoichiometric dosages. Resulting clean surfaces responded better to corrosion inhibition than did those of similar systems conditioned by natural organics, modified natural organic polymers and synthetic polyelectrolytes.

Poly(ethylene oxide) homopolymers, described in the next section, have demonstrated their value as corrosion inhibitors and synergists in suppressing the erosion-corrosion of circulating water systems, for example, open and closed cooling water systems. Hwa et al. patented this application in 1972.[17] They used alkylene oxide polymers (generally based on ethylene oxide but containing small amounts of propylene oxide) in the molecular weight range of 1 million to 7 million, finding them effective both as inhibitors and as synergists with other inhibitor agents, when used at dosages of 0.01 to 10 mg/ℓ.

H. Drag Reduction

Non-ionic poly(ethylene oxide) homopolymers ranging from 100,000 to 5 million in molecular weight are completely soluble in water up to its boiling point. They are of special interest for applications where thickening power is required. Since they are polyethers with the common structure $\text{(O-CH}_2\text{CH}_2\text{)}_n$ they hydrogen-bond strongly with water.

Polyox® WSR-301 (Union Carbide Corporation) has a molecular weight of 4 million. A 1% solution has a viscosity range of 1650 to 3850 cp at 25°C, measured by a Brookfield® viscosimeter, Spindle No. 1, 2 rpm. These Polyox® resins act to lower frictional drag by reducing turbulence. Many "wetter water" applications immediately are suggested: e.g., friction reduction in pipelines and hoses for fire fighting; storm sewers; irrigation; cooling systems; electric power storage; pipeline transmission of coal slurries; and transport of oil emulsions. They contribute to pollution control and conservation by increasing the effective carrying capacity of sewers and waste lines by reducing pumping costs and head loss. Concentrations of polymer required are quite low, falling within the range of 1 to 500 mg/ℓ by weight. Another intriguing application of Polyox® is the use to reduce frictional drag and power consumption on ships, by injecting the polymer directly into the boundary layer along the ship's surface, permitting speed increases up to 30%. Cost is a major factor here, so this method would be of most value in military marine equipment or emergency services where bursts of speed are essential. Similar considerations apply to firefighting, where the cost of polymer may be quite small in relation to the value of buildings and equipment being protected.

I. Water Conservation and Recovery

In the desalination of sea water and brackish water supplies, scale formation by calcium carbonate, calcium sulfate and magnesium hydroxide in distilling units and multiple effect evaporators is the major operating problem. Many procedures have been attempted to minimize scaling problems, and seclected polymers have shown their effectiveness in lower temperature evaporators. Polyacrylic and polymethacrylic acids and corresponding alkaline salts were patented for this service by Salutsky.[18,19] Low molecular weights showed best activity, and dosages required were in the useful range of 0.1 to 20 mg/l. The U.S. Public Health Service advised in 1967 that sodium salts of polyacrylic and polymethacrylic acids could be safely used as scale dispersants on the raw water side of evaporators producing potable water.[20]

Sexsmith et al.[21] also discussed this problem and indicated excellent scale reduction in low-pressure marine evaporators with polymaleic anhydride (producing polymaleic acid in water) in the molecular weight range of 5000 to 10,000. Dosages of 2 to 4 mg/ℓ were effective. It would be expected that acrylic and polymaleic anhydride polymers would prove equally effective in many other evaporative methods for the concentration and disposal of wastes representing significant pollution hazards.

III. LIGNIN DERIVATIVE POLYMERS

Lignin sulfonates and other modified lignin derivatives were mentioned earlier as scale-control agents, dispersants, and synergists in boiler water and cooling water treatment. Some authors hold that lignin derivatives are not polymers in the classical sense. However, all agree that lignin has a chain structure, with multiple methoxy and hydroxyl groups, and is easily modified to add sulfonic and extra OH groups, providing useful dispersant properties. Crude lignosulfonates from paper pulp mill operations are purified, then partially demethoxylated in an alkaline oxidation process which makes available multiple hydroxyl and carboxyl functional groups. Maracell® E, (Marathon Corporation) has a long history of successful use as a boiler water sludge dispersant and antifoam synergist, at dosages of 8 to 10 mg/ℓ. It is still in demand today, even though supplanted for many scale-control and dispersant operations by the new synthetic polymers. Other lignin modifications produce superior dispersants, crystal growth inhibitors, and viscosity reducers with uses in water treatment, agriculture, industrial cleaning, and numerous industrial processes. They also are useful as stabilizers for oil well drilling fluids, firefighting foams, and in mineral beneficiation processes. Other modifications exhibit strong chelation properties and so are useful in agricultural trace element addition, pesticide stabilization, and as synergists.

IV. CHELATING POLYMERS

It has long been recognized that many water-soluble polymers have complexing properties, and exhibit "threshold" dosage effects in suppressing calcium carbonate precipitation, for example, at far less than stoichiometric dosages, even though these effects may be short-term. The possibility of building actual chelating functionality into polymer structures is an intriguing approach, with many potential applications in water treatment, conservation, and pollution control. Landmark work on this concept was done by Gunderson and Kerst and the first U.S. Patents issued in 1966 and 1967.[22,23] So far, products based on these patents are not available on the market.

REFERENCES

1. **Berk, A. A.**,Questions and Answers on Boiler Feedwater Conditioning, Handbook 3, 3rd ed., U.S. Department of the Interior, Bureau of Mines, U.S. Government Printing Office, Washington, D.C., 1962.
2. **Denman, W. L. and Salutsky, M. L.** Boiler scale control with polyacrylates and polymethacrylates, in *Proc. 28th Int. Water Conf.*, Engineers Society of Western Pennsylvania, Pittsburgh, 1967, 85.
3. **Denman, W. L., Edelson, M. R, and Salutsky, M. L.**, Boiler scale control in the carbonate cycle with synthetic polymers, in *Proc. 29th Int. Water Conf.*, Engineers Society of Western Pennsylvania, Pittsburgh, 1968, 119.
4. **Dwyer, J. J. and Wilkes, J. F.**, *Pioneers in Railroad Water Treatment,* Preprints, Division of Environmental Chemistry, American Chemical Society, New York, 1976, 72.

5. **Sexsmith, D. R., Savinelli, E. A., and Beecher,J. S.**, The use of polymers for water treatment, in *Proc. 30th Int. Water Conf.*, Engineers Society of Western Pennsylvania, Pittsburgh, 1969, 15.
6. **Anon**, Betz Handbook of Industrial Water Conditioning, 6th ed., Betz Laboratories, Inc., Trerose, Pa., 1962, 161.
7. **Crum, F. B. and Wilkes, J. F.**, Sludge Conditioning and Dispersing Agents for Boiler Water Treatment, U.S. Patent 2,783,200, February 26, 1957.
8. **Obrecht, M. F., Bickel, H. H., and Wilkes, J. F.**, Waterside studies in experimental high pressure laboratory boiler, in *Proc. 15th Int. Water Conf.*, Engineers Society of Western Pennsylvania, Pittsburgh, 1954, 87.
9. **Zeleny, R. A. and Vithani, K.**, The role of organic additives in preventing scale formation on heating surfaces, *Combustion,* (3468), 47, February 1963.
10. **Denman, W. L.**,Method of Inhibiting Foam Formation in Steam Generating Systems, U.S. Patent 2,727,867, December 20, 1955.
11. **Sexsmith, D. R., Silverstein, R. M., and Savinelli, E. A.**, Polymers in cooling water treatment, in *Proc. 31st Int. Water Conf.*, Engineers Society of Western Pennsylvania, Pittsburgh, 1970, 141.
12. **Goodman, R. M.**,*Polyacrylate Anti-scalants to Promote Pollution Abatement and Maximum Reuse of Industrial Process Water,* Preprints, Division of Environmental Chemistry, American Chemical Society, San Francisco, 1976, 130.
13. **Zimmie, W. E. and Bloecher, F. W.**, Method of Removing Silt from Tanks, U. S. Patent 3,080,264, March 5, 1963.
14. **Zimmie, W. E.**, Method of Removing Tubercles from a Ferrous Surface and Inhibiting Further Tubercle Formation Thereon, U.S. Patent 3,620,667, November 16, 1971.
15. **Puckorius, P. R. and Zimmie, W. E.**, Method of Removing Tubercles Using Organic Polymers and Silica and/or Chromium Compounds, U.S. Patent 3,658,710 April 25, 1972.
16. **Hwa, C. M., Cuisia, D. G., and Gray, J. A.**, Polyacrylate-polyphosphonic Acid Treatment in Aqueous Systems, U.S. Patent 3,890,228, June 17, 1975.
17. **Hwa, C. M., Bodach, C. M., and Schroeder, C. D.**, Method of Inhibiting Corrosion in Aqueous Systems with High Molecular Weight Alkyleneoxide Polymers, U.S. Patent 3,666,404, May 30, 1972.
18. **Salutsky, M. L.**, Evaporator Saline Feedwater Treatment for Scale Control, U.S. Patent 3,444,054, May 13, 1969.
19. **Salutsky, M. L.**, Control of Scaling in Evaporators, U.S. Patent 3,514,376, May 26, 1970.
20. Private communication from Department of Health, Education, and Welfare, Public Health Service, National Center for Urban and Industrial Health, to B. M. Watson, Manager Water Process Department, W. R. Grace & Co., Research Division, Clarksville, Md., June 28, 1967.
21. **Sexsmith, D. R., Phelan, J. V., and Savinelli, E. A.**, The use of polymers for scale control in desalination evaporators, in *Proc. 32nd Int. Water Conf.*, Engineers Society of Western Pennsylvania, Pittsburgh, 1971, 19.
22. **Gunderson, L. O. and Kerst, H.**, Threshold Treatment of Water with Water-dispersible Complexing Polymers, U.S. Patent 3,285,886, November 15, 1966.
23. **Gunderson, L. O. and Kerst, H.**, Process for Inhibiting Precipitation in Water, U.S. Patent 3,331,773, August 1967.

New Developments

Chapter 15

SPECIFIC ION EXCHANGERS

C. Calmon

TABLE OF CONTENTS

I. INTRODUCTION

The volume of a feed that can be treated by a given volume of exchanger regenerated with a given dosage of regenerant depends on the ionic concentration and on the ionic selectivities. If two or three ions have nearly the same selectivities, all will be removed at the same time. Only when the selectivity values are much different can one ion be preferentially removed when in the presence of other ions. Thus it would be difficult to remove zinc from a solution containing magnesium ions, since the selectivities of the two ions are nearly the same. Barium could be readily separated from magnesium with a standard 8% DVB cross-linked sulfonated polysterene cation exchanger since their selectivity coefficients are far apart.

Frequently, it is desired to remove a specific ion (e.g., boron from irrigation waters, nitrate from drinking water) without removing other ions which are innocuous to the system in order to increase the volume of feed that can be treated with a given regenerant dosage of eluting agent.

Although the first specific ion exchanger (for potassium) was prepared in 1945, until recently, little effort was directed by the commercial manufacturers of ion exchangers to prepare specific ion exchangers. This activity is due to the awareness of the toxicity and hazards of trace quantities of certain ions entering our water environment from industrial and municipal waste effluents and the need for reducing the regenerant dosage.

II. GROUPS WITH HIGH SPECIFICITY

Usually, the ion exchangers with high specificity have polar groups which form strong complexes or tend to form precipitates with the ion in question, e. g., dipicrylamine with potassium, dimethylglyoxime with nickel, etc. Table 1 is a list of ions for which the literature gives specific or highly selective exchangers. Most of the products have been gathered from the literature and are not commercially available. The commercial suppliers of those resins that are available are also listed in Table 1. Information on the use of the ion exchangers for specific commercial applications is quite limited.

TABLE 1

Specific Groups Containing Exchangers for Specific Ions

Ion	Specific Exchanger or Group
Ammonium[1,2]	Clinoptilolite (Baroid-Division of NL)
Arsenic[3]	Fluorone
Beryllium[4]	Diallylphosphate
Bismuth[5]	Pyrogallol
Boron[6]	*N*-methylglucamine (Rohm and Haas Amberlite XE-243®)
Calcium[4,7,9]	Gallic acid, iminodiacetic acid, and diallyl phosphate
Cesium[9]	Methylene sulfonic acid (Bayer-Lewatit DN® and Diamond Shamrock Duolite C-3®)
Cobalt[7,8,10]	8-hydroxyquinoline
	B-diketone
	Ethylenediaminetetraacetic acid
Copper[10,11,12,13]	Diphenylthiourea
	8-hydroxyquinoline
	Anthranilic acid
	B-diketone
	Ethylenediaminetetraacetic acid
Cyanide[14]	Anion exchanger (Rohm and Haas Amberlite XE-275® for ferrocyanide)
Germanium[3,15]	Fluorone
	Polydric phenols
Gold[16]	Polyisothiouronium (Ayalon SRAFION-NMRR®)
Iron[4,13,17,18,19]	Alginic acid
	Diallylphosphate
	Hydroxamic acid
	M-phenylglycine
	Thiol (AKZO Chemical Co.)
	Anthranilic acid
Lead[5,20]	Pyrogallol (AKZO Chemical Co.-IMAC T-73®)
Magnesium[17,21]	Alginic acid
	Phenyldiaminoacetic acid
Mercury[5,20,22,23]	Thiol (AKZO Chemical Co-Imac TMR®; Diamond Shamrock Duolite ES464®)
	Polyisothrouronium (Ayalon SRAFION-NMRR
	Pyrogallol
	Dithiocarbamate (Nippon Soda-NISSO Alm®)
Nickel[20,24,25,26,27]	Dimethylglyoxime
	Thiol (AKZO Chemical Co.)
	8-hydroxyquinoline
	B-diketone
	Ethylenediaminetetraacetic acid
Nitrate[28]	Alkylated amidines
Palladium[29,30,31]	Aminophenol & nitro groups
	Guanidine
Platinum[16,32]	Polyisothiouronium (Ayalon SRAFION-NMRR)
	Guanidine
	Polyisothiouronium (Ayalon SRAFION-NMRR)
Potassium[33]	Dipicrylamine
Silver[34,35]	Thiol
	Amino-carboxylic
Strontium[5]	Diallyphosphate
Thorium[5]	Arsonic acid
Titanium[24]	Chromotropic acid
Uranyl oxide[36]	Schiff base — dinitrophenol
Zinc[37,38]	Anthranilic acid
	Phosphonic acid resin
Zirconium[39]	Phosphonic

REFERENCES

1. **Kepple, L. G.,** Ammonia removal and recovery becomes possible, *Water and Sewage Works,* 42, 1974.
2. **Sims, R. C. and Hindin, E.,**Division of Environmental Chemistry, 173rd Meeting American Chemical Society, March 20, 1977.
3. **Seidl, J. and Stamberg, J.,** A new type of selective ion exchanges, *Chem. Ind. (London),* p. 1190, 1960.
4. **Kennedy, J. and Davies, R. V.,** Separation of uranium from heavy metals with phosphoric acid chelating resins, *J. Appl. Chem.,* 9, 32, 1959.
5. **Smirnov, A. S. and Bluvstein, M. ,** Effect of the solubility of the sorption compounds on the sorption capacity of cationites, *Dokl. Akad. Nauk, SSSR,* 70, 449, 1950.
6. **Kunin, R. and Preuss, A. F.,** Characterization of a boron specific ion exchange resin, *Ind. Eng. Chem. Prod. Res. Dev.,* 3, 304, 1964.
7. **Parrish, I. R.,** Selective ion exchangers from polystyrene, *Chem. Ind. , (London),* p. 137, 1956.
8. **Pepper, D. W. and Hale, D. K.,** Ion exchange resins structure synthesis and general properties, in *Ion Exchange and its Applications,* Proc. Ion Exchange Conf., Society Chemical Industry, London 1955.
9. **Bayer, A. G.,** The Lewatits; and Diamond Shamrock, Duolite Ion Exchange Manual, 1955.
10. **Staudinger, H. and Haberle, M.,** Polyvinyl acetyl, acetate, *Angew Chem.,* 9, 52, 1952.
11. **Bressler, S. E.,** Development in the synthesis and application of ion exchange and electron exchange resins, *Russ. Chemi. Rev.,* 29, 469, 1960.
12. **Pennington, L.D. and Williams, M. B.,** Chelating ion exchange resins, *Ind. Eng. Chem.,* 51, 759, 1959.
13. **Gregor, H. P., Taifer, M., Citarel, L., and Becker, E. I.,** Chelate ion exchange resins, *Ind. Eng. Chem.,* 44, 2834, 1952.
14. **Avery, N. L. and Fries, W.,** Selected removal of organics from industrial waste effluents with ion exchange, *Ind. Eng. Chem. Prod. Res. Dev.,* 14, 102, 1975.
15. **Johanson, C. E.,** *Talanta,* 19, 1349, 1972.
16. **Calmon, C.,** Precious metal recovery, in *Ion Exchange For Pollution Control,* Vol. 1, CRC Press, Boca Raton, Fla., chap. 18, 1979.
17. **Specker, H. and Hartkamp, H.,** Analytical use of alginic acid as the cation exchanger, *Anal. Chem.,* 40, 167, 1953.
18. **Cornaz, I. P. and Deuel, H.,** Ion exchangers IV-selected ion exchangers for ferric ions, *Experentia,* 10, 137, 1954.
19. **Gregor, H. P., Dolar, D. and Hoschele, G. K.,** Polythiolstyrene — a new oxidation-reduction ion exchange resin, *J. Am. Chem. Soc.,* 77, 3675, 1975.
20. **Koerts, K.,** Selective removal of mercury, lead, zinc, copper and silver, American Institute Chemical Engineers/Environmental Protection Agency Meeting, Chicago, May 1975.
21. **Tagaki, T. E. and Imoto, H.,** *Jpn. Analyst,* 7, 565, 1953.
22. **Law, S. L.,** Methyl mercury and inorganic mercury collection by a selective chelating resin, *Science,* 174, 285, 1971.
23. **Calmon, C.,** Mercury removal by ion exchange, in *Ion Exchange for Pollution Control,* Vol. 1, CRC Press, Boca Raton, Fla., chap. 16, 1979.
24. **Stamberg, J., Seidl, J., and Rahm, J.,** The synthesis of a unique nickel selective ion exchanger, *J. Polymer Sci.,* 31, 15, 1958.
25. **Lillin, H.,** *Angew Chem.,* 66, 649, 1954.
26. **Cairns, T. L., Engelhardt, V. A., and Forshey, W. O.,** Synthesis and reactions of ethyl 3-ethoxy-2-4-pentadieonate, *J. Am. Chem. Soc.,* 77, 4669, 1955.
27. **Blasius, E. and Olbrich, G.,** Complexion exchange resin and analytical use, *Zh. Anal. Khim.,* 151, 81, 1956.
28. **Dow Chemical Company,** Nitrate removal from wastewater by ion exchange, Water Pollution Control Research Series Report No. 17010-FSJ01/71, Washington, D.C., January 1971.
29. **Dedkov, Y. M., Eliseeva, O. P., Ermakou, A. N., Savvin, S. B and Slotintseva, M. O.,** *Zh. Anal. Khim.,* 27, 726, 1972.
30. **Das, H. A. et al,** *Radioanal. Lett.,* 8, 257, 1971.
31. **Koster, G. and Schumckler, G.,** Separation of noble metals from base metals by means of a new chelating resin, *Anal. Chim. Acta,* 179, 38, 1967.
32. **Gulko, A., Feigenbaum , H. and Schmidt, G.,** Separation of palladium II and platinum II chlorides by means of a guarnidine resin, *Anal. Chim. Acta,* 59, 397, 1971.
33. **Skogseid, A., Thesis,** *Nor. Tek. Hogs.,* p. 12, 1948.

34. **Brechling, C. and Gartner, K.,** *Symp. on the Anomalies of Ion Exchange Reactions,* Akad Verlag, Berlin, 1962, 187.
35. **Hering, R.,** An ion exchange resin with attached nitrilotriac acid, *Prakt. Chem.,* 14, 285, 1961.
36. **Bayer, E.,** British Patent 803, 199, 1958.
37. **Chamberlain, D. G.,** Selective removal of zinc from cooling tower blowdown by ion exchange, Proc. 31st Ind. Water. Conf., Pittsburgh, p. 151, 1970.
38. **Anderson, R. E.,** The concentration of trace heavy metals with ion exchange resins, *Conf. on Toxic Metals,* Report EPA 902/9-74-001, Environmental Protection Agency, Princeton Univ., 1973.
39. **Kennedy, J., Burford, F. A., and Sammes, P. G.,** Selective adsorption of sexivalent uranium by a nonionic phosphorylated resin from solution of dibutyl phosphoric acid in benzene, *J. Inorg. Nucl. Chem.,* 14, 114, 1960.

Chapter 16

IMMOBILIZED ENZYMES

F. L. Burnett

TABLE OF CONTENTS

I. INTRODUCTION

From the beginning of time, enzymes have cured man's meat, tanned his hide, and prepared his favorite drink. It is only in the last two centuries that he discovered that special proteins (enzymes) were the responsible agents. Today enzymes are known as nature's catalysts, synthesizing, degrading, or transforming molecules at rates several orders of magnitude faster than they naturally proceed. In addition, these processes usually take place under mild conditions in an aqueous medium. Researchers have learned that enzymes are highly specific, are usually effective on only one substrate, and that they result in a single product. Thousands of enzyme-catalyzed reactions have been identified, hundreds of enzymes have been isolated, but less than twenty have commercial significance. Most of these commercial applications are in the food process industry, specifically starch liquefaction, glucose production, cheese manufacture, baking, and brewing. Many other applications for enzymes have been demonstrated, but the low cost effectiveness of current technology precludes their use. First, enzyme purification is a tedious, costly process. Second, when used in their soluble or free state, enzymes must be destroyed after only one use to prevent further changes in product composition. In addition to the high cost of using free enzymes, if the product is for human consumption there is a risk of adverse immunological response because the now inactive enzyme is still in the product.

In the past, even the most energy-intensive process was preferable to enzyme catalysis because energy was cheap. But as the energy crisis arose and researchers became aware of rapidly dwindling energy reserves, they sought other processes which would conserve energy. By taking advantage of the mild conditions of enzyme catalysis, much energy could be saved. However, destruction of the enzymes after one use seemed inefficient. This led to much research activity in immobilized enzymes.

Immobilized enzymes were described in a report by Nelson and Griffin over 50 years ago, yet that report lay dormant until the mid-1950s.[1] Even then, enzymes remained a laboratory curiosity until the late 1960s, when they were shown to have potential commercial significance. As the energy crisis deepened, the research work on immobilized enzymes increased rapidly and received national attention in the form of National Science Foundation grants through its Research Applied to Natural Needs (RANN) program. Today, the subject of immobilized enzymes is discussed in at least one international symposium each year, the most notable being the Enzyme Engineering and the Gordon Research Conferences. An easy way of keeping abreast of the rapidly changing technology in this area is by reading conference proceedings. For those who want an introduction to the field, the recently published books by Zaborsky,[2] Messing,[3] and Weetall[4] are excellent.

Why do immobilized enzymes offer so many advantages over the free enzyme? The primary reason is cost: The immobilized enzyme can be reused a number of times before it becomes ineffective. Secondly, product purity is controllable. The immobilized enzyme can be easily removed from the product; thus there is no residual enzyme action to cause unwanted side reactions. Thirdly, in products designed for human consumption, there is no danger of adverse immunological response. The enzyme detergent scare has made the Food and Drug Administration particularly sensitive to this point.

II. METHODS OF IMMOBILIZATION

Much early work was done to define methods of immobilization which would allow retention of enzyme activity both during immobilization and use. Obviously, the enzyme must be immobilized without blocking the active site. Also, the substrate must have access to the enzyme, and the product must be able to leave. Many types of support materials have been studied and comparative analyses done, with the conclusion that no universal immobilization method or support material exists. Thus, screening studies with each enzyme are necessary to determine the optimum immobilization method and support material.

Basically, there are four types of immobilization methods, as shown in Table 1. The oldest and simplest is to sorb the enzyme onto the surface of the support material. Glass, ceramics, activated carbon, ion exchange resins, porous polymers, or cellulose are types of support material used. Usually, these materials are porous and have high surface areas, up to hundreds of square meters per gram allowing large quantities of enzyme to be immobilized. Pore diameter is critically important. The pores must be large enough to allow the enzyme to enter and become immobilized and large enough to allow substrate and product to enter and leave freely. A new wrinkle on this approach is to cross-link the enzyme in place on the support with a difunctional reagent

TABLE 1

Immobilization Methods

Method	Typical support materials
Sorption	Porous polymers, activated carbon, ion exchange resins
Entrapment	Polyacrylamide, silica gel, starch
Covalent attachment	Cellulosic materials, glass
Membranes	Cellulose acetate, collagen

such as glutaraldehyde.[5] This increases the useful life of the immobilized enzyme by reducing the amount that is removed by leaching during use.

A second approach is to entrap the enzyme in a cross-linked gel. Polyacrylamide, silica gel, and starch have been studied extensively. These gels have a sufficiently high cross-link density so that the large enzyme molecule is trapped within the matrix. Nevertheless, they are porous enough to allow substrate and product to move freely. A large number of enzymes immobilized by this technique is available in research quantities from laboratory supply houses.

Covalent attachment is the most commonly used immobilization method. Generally, the support material must be activated before the enzyme can be immobilized. For example, cellulosic materials are converted to the azide, glass is silylated to the amino alkyl or aryl derivative and agarose can be activated with cyanogen bromide. This method suffers from a cost disadvantage, because the activation steps use expensive and sophisticated chemicals and technology. However, it is popular because, at first glance, it would seem to offer the greatest stability. The enzyme is chemically bonded to the support, and indeed, there is less leakage of enzymes immobilized by this technique. However, loss of activity occurs by other mechanisms as well, and reported stabilities of materials prepared by covalent attachment are no longer than those prepared by some of the other techniques. Research quantities of enzymes immobilized by this method are also commercially available.

A fourth method utilizes semipermeable ultrafiltration membranes. The success of this method depends on the use of a membrane with a molecular weight cut-off high enough to allow substrate and/or product to permeate through the membrane but low enough to retain the enzyme. Immobilization within hollow fibers, first reported by Rony in 1972,[6] is the subject of much current research.

Other miscellaneous immobilization methods include enzyme cross-linking microencapsulation, and copolymerization. Although these methods have been or are still being researched, their potential commercial significance seems small.

III. PROPERTIES

As expected, the properties of immobilized enzymes are different from those of the free enzyme. Usually, not all of the enzyme activity presented to the support appears in the immobilized enzyme. Some of the enzyme is not attached, and some is rendered inactive either by blockage of its active site or by denaturing. Retention of up to 90% of the free enzyme activity has been reported.[7] The pH of maximum activity is usually displaced from that of the free enzyme. The shift can be either higher or lower depending on the ionic nature of the support. The same Michaelis-Menten kinetic scheme used to describe the reaction rates of free enzymes is used for immobilized enzymes. The only new parameter that needs to be considered is the effect of diffusion. In general, immobilized enzymes show reaction rates slower than in the free state; some of the reduced rate is due to extended diffusion paths. Remarkably, reports of increased reaction rates have been made.[8]

While discussing kinetics, the types of reactor configurations used must be considered. Most of the immobilized enzymes are prepared in particle form similar in size to ion exchange resins. Thus, for most applications, a column or packed-bed reactor is employed. These are operated either downflow or upflow. The stirred-tank or continuous stirred-tank reactor design has been studied for some applications. Of course, for those enzymes used in membrane systems, the usual ultrafiltration apparatus would be employed. The kinetic scheme for these systems is being developed.

Stability of immobilized enzymes depends on a number of factors. Although many enzymes show increased temperature stability when immobilized, they all eventually lose activity. In most reported studies of long-term continuous use, the activity has fallen off with increasing time due not to loss of enzyme, but to denaturation.[8] A convenient measure of stability is half-life, the length of time required for the activity to decrease to one half its initial value. The choice of optimum operating conditions is based primarily on an economic evaluation. Because the enzyme thermally denatures, lower operating temperatures usually result in increased half-lives. However, reaction rates usually decrease with temperature, so a compromise temperature which results in the lowest operating cost consistent with high productivity is chosen.

IV. APPLICATIONS

Only a few of the potential applications for immobilized enzymes have been commercialized. Some are shown in Table 2. In 1969, Tosa et al. in Japan developed the first commercial immobilized enzyme process — the separation of L-amino acids from their DL-amino acid mixtures.[9] There are now several similar applications in Japan. In the U.S., the food industry early recognized the value of immobilized enzymes, and in fact, the corn starch industry has the only existing large-scale application — the conversion of glucose (dextrose) to fructose. The enzyme responsible for this conversion, glucose isomerase, is difficult to obtain; however, with wildly fluctuating sugar prices, the growth of this product is expected to be phenomenal. There are at least four manufacturers offering systems containing immobilized glucose isomerase.

Other applications in the food industry are on the verge of becoming commercialized. The cornstarch industry is using immobilized glucoamylase to degrade cornstarch into glucose; a pilot unit has been operated at Iowa State University in Ames, since 1975.[10] The dairy foods industry uses immobilized lactase to treat milk or cheese whey, converting lactose into glucose and galactose. Thus, a high BOD and COD pollutant becomes a useful sweetener. The conversion is done after removing the protein from the whey by ultrafiltration.

Potential applications in pollution control have been demonstrated. Immobilized α-amylase has been shown to hydrolyze starch in the "white water" from the forest product industry. [11] Instead of discarding this water as waste, it could be recycled. Another application is the use of immobilized cellulose to convert cellulose from paper wastes into glucose.[12] The product can be used as either a food or a chemical feedstock source. This could become a significant application as food and energy resources dwin-

TABLE 2

Current and Potential Applications of Immobilized Enzymes

Enzyme	Application
Pharmaceutical	
Amino acid acylase	Separation of DL-amino acids
Food Processing	
Gluco-amylase	Conversion of starch to glucose
Glucose isomerase	Conversion of glucose to fructose
Papain	Beer chillproofing
α-Amylase	Waste water treatment from paper industry
Lactase	Lactose hydrolysis in milk or whey
Clinical	
Glucose oxidase/catalase	Glucose analysis

dle. Two other applications were described at a waste water treatment symposium held in 1974. The first was the treatment of urinous waste water with urease immobilized on collagen chips. Although the rate of hydrolysis was too slow, it was suggested that membrane-immobilized urease would improve the kinetics.[13] In the other paper, immobilized enzymes were used to disinfect viruses. Although waterborne viruses were disinfected at a slow rate, airborne viruses were disinfected at a high rate.[14] The obvious application for the treatment of sewage waste has been largely discounted. Sewage treatment requires a wide variety of enzymes each needing different conditions for maximum activity. Also, the potential for fouling the porous support is high.

Perhaps the greatest activity in applications research is taking place in the analytical and clinical testing fields. Already there are no less than a dozen enzyme electrodes, which because of enzyme specificity promise extreme accuracy.[15] A rapid increase in the number of medical diagnostic tests has led to the development of metabolite-specific analyzers containing immobilized enzymes. The first commercialized unit is a glucose analyzer.* Future applications include the use of immobilized enzymes as extracorporeal shunts, either to supply a missing enzyme function or remove a toxic chemical from body fluids, e.g., blood stream.

V. FUTURE DEVELOPMENTS

It is not yet clear whether or not processing with immobilized enzymes will become a unit operation similar to ion exchange. Many potential applications which look interesting on paper will probably never be economically viable. Nature's catalysts are so efficient that she has not had to concentrate any one enzyme in a single organism. Thus, enzymes are expensive because of difficulties in isolation and purification. Enzyme costs will decrease, however, as genetic engineers learn how to tailor-make organisms which provide high concentrations of specific enzymes.

There is a market for immobilized enzymes now. Those processes already employing enzymes will most certainly be able to use immobilized enzymes. The cornstarch hydrolysate and the clinical testing industries are prime examples of successful uses of immobilized enzymes. Also, there will be those special applications where enzyme specificity will give better product quality at higher yields or where the conversion cannot be done any other way. One potential application of interest to membrane chemists is the degradation of the high-molecular-weight materials which foul membranes. A hydrolase enzyme which will break down the foulants into smaller molecules allowing them to pass through the membrane instead of remaining on the surface undoubtedly exists for each of these materials. It is conceivable that the enzymes themselves could be immobilized on the membrane surface by one of a number of techniques (see Reference 2).

The immobilized enzyme field is in its infancy. It will be several years before enzyme costs decrease to a level where the immobilized enzyme unit process is economically competitive with conventional processes. However, a major shortage in energy resources could lead to early widespread acceptance of the process. Despite what happens on the large commercial scale, there will always be a small market for applications which take advantage of enzyme specificity to convert materials in ways that cannot be done conventionally.

* The immobilized enzyme Catalinks® is available from Miles Laboratories, Inc.

REFERENCES

1. **Nelson, J. M. and Griffin, E. G.**, Adsorption of invertase, *J. Am. Chem. Soc.*, 38, 1109, 1916.
2. **Zaborsky, O. R.**,*Immobilized Enzymes,* CRC Press, Cleveland, Ohio, 1973.
3. **Messing, R. A., Ed.**, *Immobilized Enzymes for Industrial Reactors,* Academic Press, New York, 1975.
4. **Weetall, H. H., Ed.**, *Immobilized Enzymes, Antigens, Antibodies and Peptides,* Marcel Dekker, New York, 1975.
5. **Stanley, W. L. and Olson, A. C.**, Preparation of insolubilized enzymes, U.S. Patent 3,767,531, 1973.
6. **Rony, P. R.**, Hollow fiber enzyme reactors, *J. Am. Chem. Soc.*, 94, 8247, 1972.
7. **Tosa, T., Mori, T., Fuse, N., and Chibata, I.**, Studies on continuous enzyme reactions. I. Screening of carriers for preparation of water-insoluble aminoacylase, *Enzymologia,* 31, 214, 1966.
8. **Sharp, A. K., Kay, G., and Lilly, M. D.**, The kinetics of β-galactosidase attached to porous cellulose, *Biotechnol. Bioeng.*, 11, 363, 1969.
9. **Weetall, H. H.**, Applications of immobilized enzymes, in *Immobilized Enzymes for Industrial Reactors,* Messing, R. A., Ed., Academic Press, New York, 1975, 214.
10. **Weetall, H. H.**, Applications of Immobilized enzymes, in *Immobilized Enzymes for Industrial Reactions,* Messing, R. A., Ed., Academic Press, New York, 1975, 207.
11. **Smiley, K. L., Boundy, J. A., Hofreiter, B. T., and Rogivin, S. P.**, Immobilized α-amaylase for clarification of colloidal starch-clay suspensions, in *Immobilized Enzymes in Food and Microbial Processes,* Olson, A. C. and Cooney, C. L., Eds., Plenum Press, New York, 1974.
12. **Mandels, M. and Kostich, J.**, Enzymatic hydrolysis of cellulose to soluble gases, U.S. Patent 3,764,475, 1973.
13. **Davidson, B., Vieth, W. R., Wang, S. S., Zwiebel, S., and Gilmore, R.**, Enzymes and wastewater treatment, in *Water-1974.* I. Industrial Water Treatment, Bennett, G. F., Ed., American Institute of Chemical Engineers, New York, 1974.
14. **Enright, J. T. and Kirwan, D. J.**, Disinfection of viruses in wastewater using immobilized enzymes, in *Water-1974.* I. Industrial Water Treatment, Bennett, G. F., Ed., American Institute of Chemical Engineers, New York, 1974.
15. **Guilbault, G. G.**, Enzyme electrode probes, in *Immobilized Enzymes, Antigens, Antibodies and Peptides,* Weetall, H. H., Ed., Marcel Dekker, New York, 1975.

Chapter 17

LIQUID ION EXCHANGERS

C. Calmon

TABLE OF CONTENTS

I. INTRODUCTION

Liquid organic polar molecules which are insoluble in water can act as cation or anion exchangers, depending on the polar group.[1,2] Liquid anion exchangers contain either primary, secondary, tertiary, or quaternary amines, for example, trialkylmethylamine, dilaurylamine, methyloctylamine, and tetrabutylammonium, respectively. The liquid cation exchangers have as their polar group either the sulfonic (SO_3H), carboxylic (COOH), phosphoric, phosphonic $PO(OH)_2$, or phosphorous (POOH) acid.

II. PROCESS STEPS AND APPLICATIONS

The steps involved in their application consist of the following: (1) the liquid ion exchanger is dissolved in a solvent immiscible in water, (2) the aqueous solution containing the ion to be removed is contacted with the above organic solution so that exchange between the ion in the aqueous phase and the counter-ion of the organic exchanger will take place, (3) the two phases are separated, and (4) the organic mixture is then contacted with an acid or alkali to elute or strip the ions to be recovered and to convert the liquid ion exchanger to its initial form.

The solvent for the liquid ion exchanger is very important as it affects the pK of the liquid ion exchanger which is important in the efficiency of the extraction. Some of the solvents used are benzene, carbon tetrachloride, cyclohexane, kerosene, toluene, n-heptane, octane, petroleum ether, xylene, etc. While both the liquid ion exchanger and solvent are important in determining the selectivity of an ion, the selectivity may be improved or increased with the addition of neutral molecules (e.g., tri-n-butyl phosphate).

Liquid ion exchangers can be used for extracting dissociated species and then stripping these with concentrated solution. Some applications found feasible are

1. Desalination of brackish waters with a secondary amine as the anion exchanger.

2. Seawater softening with an amine in the bicarbonate form — the calcium bicarbonate formed can be reacted with calcium hydroxide to form a precipitate of $CaCO_3$.
3. Phenol extraction with a secondary amine — distillation is used to accomplish the stripping of the phenol.
4 Fission products can be recovered from waste solutions.
5. Extraction of acids or bases, i.e., conversion of a solution to a salt form.

III. PROCESS ADVANTAGES

Because of the higher content of polar groups per unit weight or volume, liquid ion exchangers have higher capacities than polymeric exchangers. Continuous systems can be easily devised. The elution efficiency is higher because diffusion is not important. Other advantages are low rinse requirements, fast kinetics, and more concentrated solutions that can be obtained than with standard ion exchangers.[3]

IV. PROCESS LIMITATIONS

The process limitations are some loss of the liquid ion exchanger due to trace solubilities and adherence to solids. This represents an economic loss. In addition, in many instances the traces remaining in the aqueous phase become pollutants. Frequently, the mixture formed in step 2 is in an emulsion form so that separation into the phases of step 3 is difficult.[3]

REFERENCES

1. **Marcus, Y. and Kertes, A S.,** *Ion Exchange and Solvent Extraction,* Interscience, New York, 1966.
2. **Hogfeldt, E.,** Liquid ion exchangers, in *Ion Exchange,* Vol. 1, Marinsky, S., Ed., Marcel Dekker, New York, 1966, 139.
3. **Lewis, C. J.,** Liquid-liquid Extraction for Ion Exchange Recovery or Treatment of Industrial Wastes, 12th Industrial Waste Conf., Lafayette, Ind., 1957.

Chapter 18

LIQUID MEMBRANE SYSTEMS

J. W. Frankenfeld and N. N. Li

TABLE OF CONTENTS

I. INTRODUCTION

The recent discovery of liquid membranes[1,2] has greatly extended the application of liquid ion exchange to the extraction of metals and other substances from aqueous systems. Oil-soluble liquid ion-exchange (LIX)® reagents, themselves a fairly recent development,[3] are ideally suited for incorporation into liquid membrane films. Two of the most important applications for such membranes are in the fields of extractive hydrometallurgy and wastewater treatment. The characteristics of liquid ion-exchange reagents and their use in hydrometallurgy has been reviewed by Lewis.[3] This chapter will deal with the application of LIX reagents in liquid membranes to the treatment of waste water.

II. LIQUID MEMBRANE CHARACTERISTICS

Liquid membranes were invented by Li in 1968[1,2] and have been proposed as solutions to a number of waste disposal problems.[4-6] There are two major types of liquid membrane systems: a water-immiscible emulsion dispersed in water and an oil-immiscible emulsion dispersed in oil. For wastewater treatment and mineral concentration from mine leaching solutions, the former is used. The water-immiscible emulsion consists of an oil phase composed of surfactants and various additives in a hydrocarbon solvent which encapsulates microscopic droplets of an aqueous solution of appropriate reagents for removing (stripping) and trapping waste water contaminants or minerals from mine leachate. A conceptualized drawing of a liquid membrane "capsule" is given in Figure 1. The example shown is that of removal of Cu^{+2} ion according to the equations[7,8]

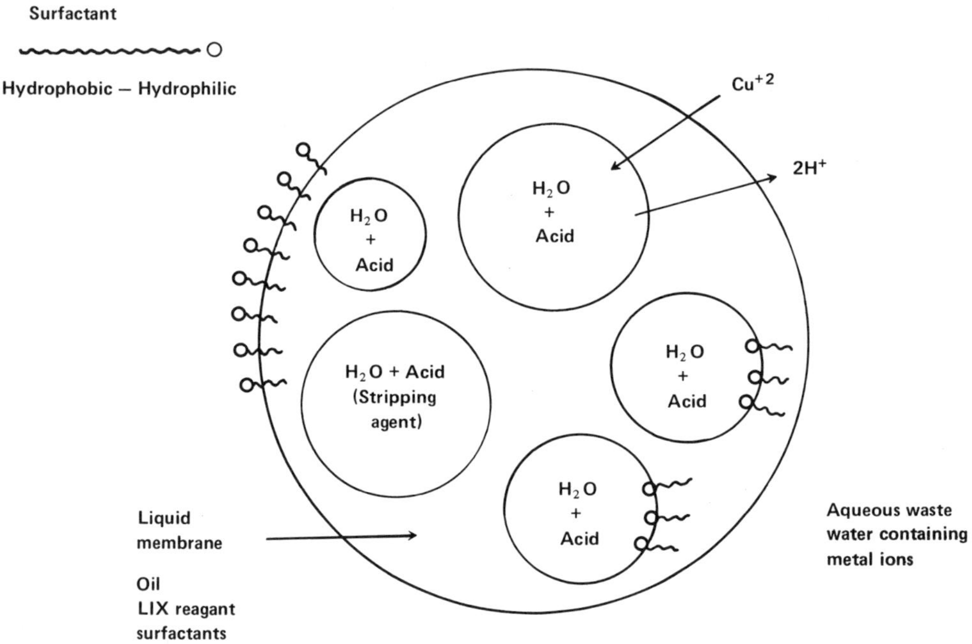

FIGURE 1. Schematic diagram of liquid membrane capsule for Cu^{+2} extraction.

Extraction:

$$\underset{\text{Org}}{2\,RH} + \underset{\text{Aq}}{Cu^{++}} \rightleftarrows \underset{\text{Org}}{R_2Cu} + \underset{\text{Aq}}{2H^+} \quad (1)$$

Stripping:

$$\underset{\text{Aq}}{2H^+} + \underset{\text{Org}}{R_2Cu} \rightleftarrows \underset{\text{Aq}}{Cu^{++}} + \underset{\text{Org}}{2RH} \quad (2)$$

where RH represents a liquid ion-exchange agent. Among the most useful for copper extraction are the oxime type produced by General Mills.[7,9]

Extraction (Equation 1) occurs at the membrane-external aqueous phase interface, while stripping (Equation 2) occurs at the membrane-internal aqueous phase interface. The overall reaction represents an exchange of a copper ion for two hydrogen ions. The copper is effectively trapped in the interior of the liquid membrane by the large excess of hydrogen ions. The effectiveness of this process for removal of copper from wastewater is discussed later in this chapter.

The above is a good example of one of the two types of "facilitated transport" through liquid membranes.[10] That is, the transport of an oil-insoluble ion is facilitated by incorporating a carrier such as a liquid ion-exchange agent in the membrane. In the cases of removal of oil-soluble contaminants such as ammonia,[8,13] organic acids,[5] and phenol,[11] no such additive is necessary. The mechanism of unassisted transport through liquid membranes has been discussed by Cahn and Li[11] and Li and Shrier.[5] This discussion is restricted to the removal of metals from wastewater by facilitated transport using liquid ion-exchange agents.

Liquid membrane processes are not limited to metal cations as shown above. Metal anions and complex anions are also removed as well by the proper choice of liquid ion-exchange and stripping agents.

III. APPLICATIONS

A. Removal of Anions from Wastewater Streams

Many anions, difficult to remove by other methods, can be extracted from wastewater by liquid membrane-liquid ion-exchange processes. A typical example is phosphate ion, which can be scavenged from aqueous systems with oil-soluble amines or quaternary ammonium salts in liquid membrane films according to the equations[5,12]

Extraction:

$$\underset{\text{Org}}{3\,[R_3N{-}HCl]} + \underset{\text{Aq}}{PO_4^{-3}} \rightleftarrows \underset{\text{Org}}{[R_3N{-}H]_3PO_4} + \underset{\text{Aq}}{3Cl} \qquad (3)$$

Stripping:

$$\underset{\text{Org}}{2[R_3N{-}H]_3PO_4} + \underset{\text{Aq}}{3CaCl_2} \rightleftarrows \underset{\text{Org}}{6\,[R_3N{-}HCl]} + \underset{\text{Aq}}{Ca_3\,(PO_4)} \qquad (4)$$

Again, extraction occurs at the membrane-external phase interface, while stripping is accomplished in the central, aqueous phase of the membrane capsules. The phosphate is effectively trapped by conversion to insoluble $Ca_3(PO_4)_3$, which can neither permeate back through the oil film nor be picked up by the amine salt. The overall reaction is exchange of phosphate and chloride ions. Counter-ions other than chloride can, of course, be used if desired. Removal rates in a typical batch experiment are shown in Table 1.[13] Nitrate ions, another contaminant difficult to remove from aqueous systems by purely chemical means, were similarly extracted.[13] These systems have the capability of removing several anionic contaminants at once. In addition, Li and Shrier have demonstrated the simultaneous extraction of organic and inorganic waste materials with a single liquid membrane formulation.[5]

B. Chromate Removal

Chromate removal from wastewater is a serious problem both in the U.S. and elsewhere. The Takuma Co., Ltd. of Japan, in conjunction with Exxon Research and

TABLE 1

Extraction of Phosphates from Waste Waters with the Liquid Membrane-liquid Ion Exchange System[13]

Contact time (min)	Wt % PO_4^{-3} in Waste Stream	
	Run No. 1[a]	Run No. 2[b]
0	0.565	0.273
2	0.265	0.123
5	0.200	0.073
18	0.050	0.016
44	—	0.004

[a] $CaCl_2$ plus NH_4OH as internal stripping agent.

[b] $CaCl_2$ plus $Ca(OH)_2$ as internal stripping agent.

Engineering Company, have recently reported the use of liquid membrane-liquid ion-exchange processes for cleanup of metal plating baths containing chromates.[8,12,22] Chromates are extracted from aqueous systems by employing either tertiary amines or quaternary ammonium salts:[8,14]

Extraction by Amine Neutralization:

$$\underset{\text{Org}}{[2R_3N]} + \underset{\text{Aq}}{[2H^+ + Cr_2O_7^{=}]} \rightleftarrows \underset{\text{Org}}{[(R_3NH)_2Cr_2O_7]} \qquad (5)$$

Extraction by Salt Formation:

$$\underset{\text{Org}}{[(R_3NH)_2\ X]} + \underset{\text{Aq}}{[2H^+ + Cr_2O_7^{-}]} \rightleftarrows \underset{\text{Org}}{[(R_3NH)_2\ Cr_2O_7]} + \underset{\text{Aq}}{[2H^+ + X^{=}]}$$

$$X = SO_4^{-2} \text{ or } Cl_2^{-} \qquad (6)$$

The exact extraction mechanism depends upon the stripping agent employed. Stripping of the chromate ion is accomplished in the aqueous, internal phase of the liquid membrane capsules by one of two methods as shown in Equations 7 and 8:

Basic Stripping:

$$\underset{\text{Org}}{[(R_3NH)_2Cr_2O_7]} + \underset{\text{Aq}}{[4(Na^+ + \bar{O}H)]} \rightleftarrows \underset{\text{Org}}{[R_3N]}$$

$$+ \underset{\text{Aq}}{[2(2Na^+ + CrO_4^{=})]} + \underset{\text{Aq}}{[3H_2O]} \qquad (7)$$

Acid Stripping:

$$\underset{\text{Org}}{[(R_3NH)_2Cr_2O_7]} + H_2SO_4 \rightleftarrows$$

$$\underset{\text{Org}}{[(R_3NH)_2SO_4]} + \underset{\text{Aq}}{[2H^+ + Cr_2O_7^{=}]} \qquad (8)$$

Stripping with a base regenerates the free amine in the membrane which then reextracts additional $Cr_2O_7^{-2}$ by Equation 5. Acid stripping regenerates the amine salt for extraction, as shown in Equation 6. Both methods have been used successfully in liquid membranes. One of the important advantages of the liquid membrane process lies in the extraction and stripping in a single stage rather than the two separate stages required by solvent extraction. In addition, by simultaneously extracting and stripping, the liquid membrane process drives the equilibrium of extraction as shown in Equations 5 and 6 to the right by trapping the complexed ions as formed. This removes the equilibrium limitation inherent in some solvent extraction methods.

Some of the typical results of laboratory experiments are given in Figure 2 and Table 2.[8,12] Figure 2 (Curve 2) shows that in a batch run nearly complete chromium removal was achieved in 10 min with a feed containing an initial chromium concentration of 400 ppm. The results of experiments at two different external pH values are given in Table 2. These data indicate that in order to transfer chromium from the external feed to the internal reagent phase of the liquid membrane emulsion, the pH of these two phases must be controlled to within certain critical limits. At constant pH of the internal phase, given by the 1% NaOH aqueous solution, the transfer of chromium was very fast when the external phase was acidic. The Cr^{+6} content of the external phase was reduced from 390 ppm to 12 ppm (97% removal) in 15 min under these conditions. However, under basic conditions, no transfer of chromium was observed.[8]

C. Separation of Heavy Metal Ions and Metal Ion Complexes

Heavy metal ions may be extracted from wastewater streams either as cations or as anionic complexes by employing liquid membrane systems containing various liquid ion-exchange agents. The removal of cations was illustrated, in the case of copper, by Equations 1 and 2. Frequently, heavy metal ions are present in the form of soluble anionic complexes with halogens or other anions. Solvent extractions of these complexes using liquid ion exchangers have been reported.[3,15-17] For example, Moore, has

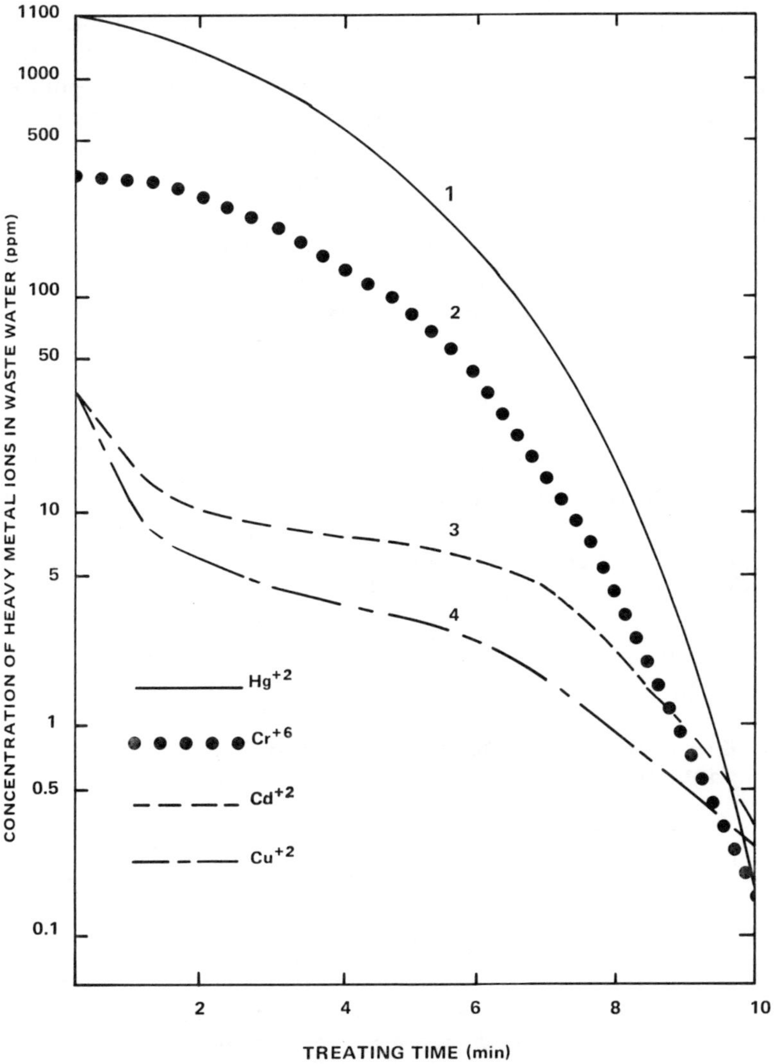

FIGURE 2. Removal of heavy metal ions from wastewater streams by liquid membrane-liquid ion-exchange technique. (Reprinted with permission from Kitagawa, T., Nishikawa, Y., Frankenfeld, J. W., and Li, N. N., *Environ. Sci. Technol.*, 11, 602, 1977. Copyright by the American Chemical Society.)

extracted mercury-halogen complexes from brine solutions.[15,16] In such solutions, mercury is present as the soluble complex:

$$Hg^{+2} + 4Cl^- \rightleftarrows Hg(Cl_4)^{-2} \quad (9)$$

Extraction is accomplished by contacting the aqueous solution with a high molecular weight amine or quaternary ammonium salt in a suitable organic diluent:

$$\underset{Org}{2\,(R_4NX)} + \underset{Aq}{HgCl_4^{-2}} \rightleftarrows \underset{Org}{(R_4N)_2HgCl_4} + \underset{Aq}{2X^-} \quad (10)$$

Regenerative stripping may be accomplished in a second step by contacting with strong acids, inorganic bases, or organic bases such as ethylenediamine or propylenediamine.[16] High molecular weight primary, secondary, and tertiary amines are effective extractants for acidic brines but quaternary ammonium salts must be used at high pH.

Extractants of the type just described have been successfully incorporated into liquid membranes for extraction of Cu^{+2} and Hg^{+2}.[8,12] The liquid membrane technique system combines extraction and stripping into a single step and numerous advantages accrue. Some typical results of batch extraction experiments are also shown in Figure 2. (Curves 2 and 4). The Hg^{+2} concentration was rapidly reduced from 1100 ppm to less than 0.2 ppm, while Cu^{+2} was reduced from 50 ppm to less than 0.3 ppm.

The removal of cadmium from metal plating baths by a liquid membrane-liquid ion-exchange technique has also been reported. [8,12] Although the general principle is the same as for Hg^{+2}, the removal of Cd^{+2} presents a somewhat different problem.[8] In the effluent from metal plating baths, cadmium exists as the tightly bound cyanide complex

$$Cd^{+2} + 4\,CN^- \rightleftarrows Cd\,(CN)_4^{-2} \quad (11)$$

This complex is readily extracted from aqueous solutions by liquid ion exchange agents such as Aliquat® 336 (methyltrioctylammonium chloride).[17] However, stripping is quite difficult because of the great stability of the $Cd(CN)_4^{-2}$ complex under basic conditions ($K_{ST} = 10^{19}$).[18] The usual stripping agents, salts and mineral acids, are ineffective even at high concentrations. Ethylenediaminetetraacetic acid (EDTA) is known

TABLE 2

Effect of External pH on Chromium Removal from Waste Water by Liquid Membrane Emulsions[a]

pH of external phase		Contact time (min.) 0	5	10	15	20	30
3.5	ppm Cr^{+6}	390	250	80	12	3	2
7.0	ppm Cr^{+6}	390		(no removal)			390

[a] Conditions: reagent phase = 2% NaOH; feed = 0.1% $K_2Cr_4O_7$; membrane/reagent weight ratio = 1.2; feed/emulsion = 3.1; temperature = 25°C.[8]

Reprinted with permission from Kitagawa, T., Nishikawa, Y., Frankenfeld, J. W. and Li, N. N., *Environ. Sci. Technol.*, 11, 602, 1977. Copyright by the American Chemical Society.

to chelate strongly with cadmium complexes.[19] In the form of its disodium salt, EDTA successfully removed cadmium from Aliquat 336 with regeneration of the ion carrier so long as the pH of the internal phase of the emulsion was maintained in the range of 4 to 6.

This procedure was studied by Kitagawa and the Exxon group,[8,12] who report the results shown in Figure 2 (Curve 3). The cadmium content of the actual plating bath effluent was rapidly reduced from 50 ppm to less than 0.5 ppm.

D. Use of Biological Systems as Trapping Agents

Mohan and Li[6,20,21] reported a novel use of biological enzyme systems to trap nitrates and nitrites in liquid membrane formulations. This process has promise for large-scale continuous operations of secondary and tertiary waste water treatment. Various amines were used as liquid ion-exchange agents to transfer the NO_2^- or NO_3^- across the membrane film. Particularly effective was Amberlite® LA-2, an oil-soluble secondary amine manufactured by Rohm and Haas. This amine readily extracted the anions from weakly acidic solutions. In one modification of the process, stripping in the internal phase of the liquid membrane emulsions was accomplished by buffered salt solutions of either purified enzymes or cell-free extracts from *Micrococcus denitrificans.*[6] The extracts rapidly reduced the nitrate or nitrite to products which were not reextracted by the liquid ion-exchange agent. Thus, the reduction products were efficiently trapped within the membrane. Nitrite was shown to be an intermediate product in the reduction of nitrate. The presumed final product was elemental nitrogen.[6]

This concept was carried one step further by encapsulating intact cells of *M. denitrificans* in liquid membrane emulsions.[20,21] Again, amines were used as transferring agents. In these experiments, the disappearance of nitrate and the formation of nitrite were measured. Both NO_2^- and NO_3^- were readily reduced, probably to N_2. The rapid reduction of NO_3^- achieved is illustrated in Figure 3.[20] As NO_3^- was reduced in the internal phase of the membrane NO_2^- was liberated and was carried to the external phase where its appearance was monitored (see Figure 3). However, the rate of NO_2^- formation was much slower than NO_3^- reduction indicating most of the nitrite was reduced further as formed. This sequential reduction is illustrated by the data in Table 3. The NO_3^- was completely reduced in 120 min contacting, while 90% of the NO_2^- formed was reduced in the same period.

The encapsulated cells remained intact and active for periods of at least 5 days. It was claimed that this period could be extended even further by encapsulating special nutrients along with the cells.[20] In addition, such known enzyme inhibitors as $HgCl_2$ in the external phase had no effect on the encapsulated cells, indicating complete protection by the surrounding membrane. Nonencapsulated cells were inactive in the same medium. This has considerable importance in developing wastewater treatment processes since many effluents to be treated contain such inhibitors.[20,21] Mohan and Li also showed that the process could be carried out continuously. A special apparatus was designed and has been described.[20]

IV. PROCESS AND ECONOMIC CONSIDERATIONS

The Takuma Company of Japan has evaluated liquid membrane processes for heavy metals removal in a continuous, two-stage counter-current pilot plant and reported the results in two recent publications.[8,12] These studies included the scavenging of NH_4^+ as well as Cr^{+6}, Hg^{+2}, Cd^{+2}, and Cu^{+2} by processes similar to those described above. The pilot plant equipment was also detailed.[12] Typical results, illustrated by the removal of Cr^{+6} from the effluent of a metal plating process are shown in Table 4. In

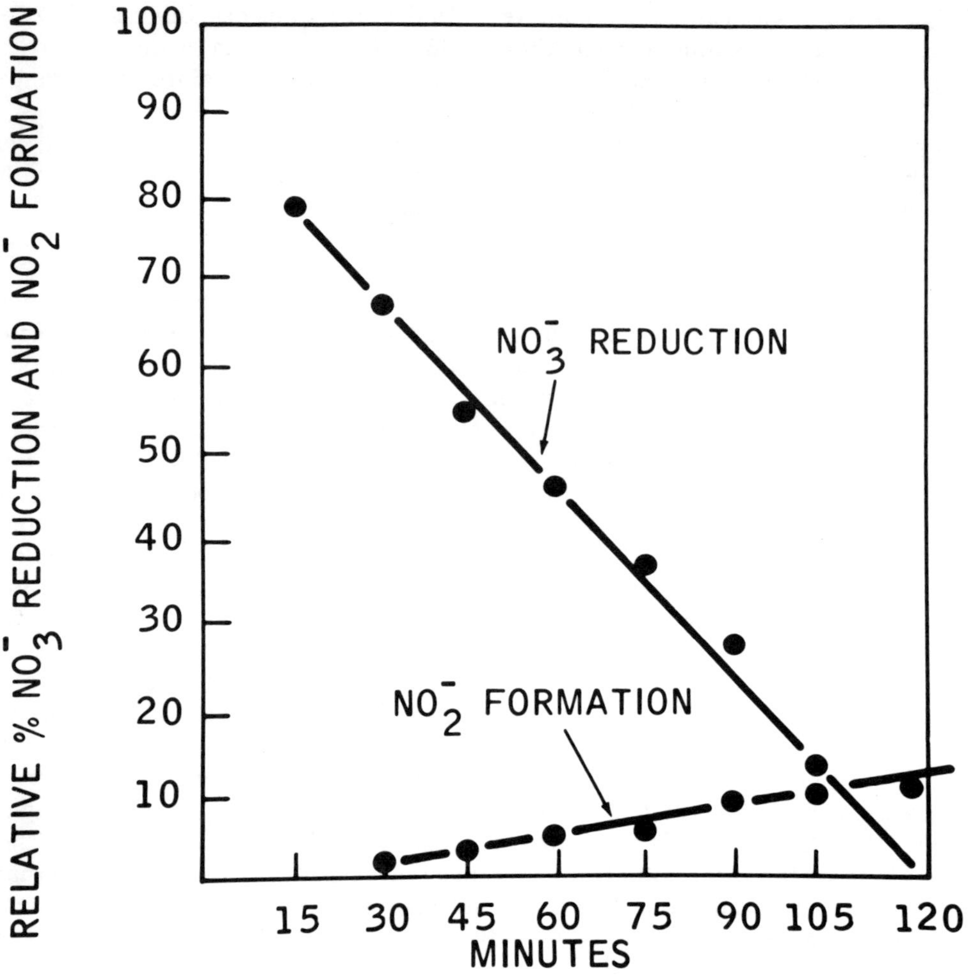

FIGURE 3. Reduction of NO_3^- and formation of NO_2^- by liquid membrane encapsulated whole viable cells of *Micrococcus denitrificans* ATCC 21909. (From Mohan, R. R. and Li, N. N., *Biotechnol. Bioeng.*, 17, 1137, 1975. With permission.)

TABLE 3

Simultaneous Reduction of NO_3^- and NO_2^- by Encapsulated Viable Whole Cells of Micrococcus denitrificans ATCC 21909

Time (min)	NO_3^- Reduced (%)	NO_2^- in aqueous phase (%)	NO_3^- Reduced beyond NO_2^- (%)
0	0	0	0
15	20	0	20
30	32	1.6	30
45	45	3.2	42
60	54	4.8	49
75	63	5.8	57
90	73	8.3	65
105	88	9.2	79
120	100	10.0	90

From Mohan, R. R. and Li, N. N., *Biotechnol. Bioeng.*, 17, 1137, 1975. With permission.

all cases, the concentration of Cr^{+6} in the wastewater was reduced, under continuous flow conditions, from 100 ppm to less than 1 ppm.

Takuma estimates that the treatment cost of Cr^{+6} cleanup will be about 15¢/m³ of wastewater (57¢/1000 gal) for a plant with a 100 m³/hr capacity and an emulsion recycling capability. Plant cost, installation, chemicals, personal expenses, and other necessary expenditures were included in this estimate.[12] This cost is quite attractive compared to most industrial wastewater cleanup processes. Furthermore, there are no economically feasible alternatives for achieving the low levels of metal required by Japanese law.

This new concept, the combination of liquid membrane and liquid ion-exchange techniques, is just emerging as a potentially useful processing technique. It would appear that such processes may play an important role in solving future wastewater treatment problems.

TABLE 4

Removal of Cr^{+6} by Two-stage Pilot Plant

Test no.	1	2	3
Feed to test unit	Wastewater containing 100 ppm Cr^{+6} from plating process		
Liquid membrane emulsion			
Reagent solution	10% NaOH	20% NaOH	20% NaOH
Ratio membrane/reagent solution	1/2	1/2	1/2
Flow rates (cc/min)			
Wastewater	300	300	300
Emulsion	280	105	64
Treat ratio	1/4.85	1/4.85	1/3.40
Mixing speed (rpm)	200	200	300
Concentration of Cr^{+6} after treatment			
In effluent	1 ppm	1 ppm	1 ppm
In emulsion	35,800 ppm	108,000 ppm	182,000 ppm

Reprinted with permission from Kitagawa, T., Nishikawa, Y., Frankenfeld, J. W., and Li, N. N., *Environ. Sci. Technol.*, 11, 602, 1977. Copyright by The American Chemical Society.

REFERENCES

1. **Li, N. N.**, U. S. Patent 3,410,794, 1968.
2. **Li, N. N.**, Permeation through liquid surfactant membranes, *AICHE J.*, 17, 459, 1971.
3. **Lewis, C. J.**, *Recent Developments in Separation Science*, Vol. 2, Li, N. N., Ed., CRC Press, Cleveland, Ohio, 1972, 47.
4. **Li, N. N., Cahn, R. P., and Shrier, A. L.**, U.S. Patent 3,779,907, 1973.
5. **Li, N N. and Shrier, A.L.**, *Recent Developments in Separation Science*, Vol. 1, Li, N. N., Ed., CRC Press, Cleveland, Ohio, 1972, p. 163.
6. **Mohan, R. R. and Li, N. N.**, Reduction and separation of nitrate and nitrite by liquid membrane-encapsulated enzymes, *Biotechnol. Bioeng.*, 16, 513, 1974.
7. **Merigold, C. R. and House, J. E.**, The Application of Liquid Ion Exchange Technology to the Recovery of Copper, presented at Copper Technology Seminar, U.S. Department of Commerce, Bureau of East-West Trade, Washington, D. C., December 1975.
8. **Kitagawa, T., Nishikawa, Y., Frankenfeld, J. W., and Li, N. N.**, Waste Water Treatment by a Liquid Membrane Process, *Environ. Sci. Technol.*, 11, 602, 1977.
9. **Kordosky, G. A., MacKay, K. D., and Virnig, M. J.**, A New Generation Copper Extractant, presented at American Institute of Mining, Metallurgical, and Petroleum Engineers (AIME) Annu. Meet., Las Vegas, Nevada, February 22 to 26, 1976.
10. **Matulevicius, E. S. and Li, N. N.**, Facilitation Transport Through Liquid Membranes, presented at American Institute of Chemical Engineers Nat. Meet., Pittsburgh, Pa., June 3 to 5, 1974.
11. **Cahn, R. P. and Li, N. N.**, Separation of phenol from waste water by the liquid membrane technique, *Sep. Sci.*, 9, 505, 1974.
12. **Kitagawa, T. and Nishikawa, N.**, Waste Water Treatment by Liquid Membrane Process, presented at American Chemical Society Centennial Symp. Separation and Encapsulation by Liquid Membranes, New York, April 6, 1976.
13. **Li, N. N., Cahn, R. P., and Shrier, A. L.**, U. S. Patent 3,779,907, 1973
14. General Mills Corporation, Chromium, Technical Bulletin CDSI-61, Minneapolis, 1961.
15. **Moore, F. L.**, Liquid-liquid extractions of mercury with high molecular weight amines from iodide and bromide solutions, *Sep. Sci.*, 7, 505, 1972.
16. **Moore, F. L.**, Solvent extraction of mercury from brine solutions with high molecular weight amines, *Environ. Sci. Techol.*, 6, 525, 1972.
17. **McDonald, C. W. and Moore, F. L.**, Liquid-liquid extraction of cadmium with high molecular weight amines from iodide solutions, *Anal. Chem.*, 45, 983, 1973; Aliquat® 336 is a registered trade mark of General Mills Corporation.
18. **Cotton, F. A. and Wilkinson, G.**, *Advanced Inorganic Chemistry*, 3rd ed., Interscience, New York, 1966, 514.
19. **Irving, H.M.N.H. and Al-Jariah, R.H.**, *Anal. Chim. Acta*, 74, 321, 1975.
20. **Mohan, R. R. and Li, N. N**, Nitrate and nitrite reduction by liquid membrane encapsulated whole cells, *Biotechnol. Bioeng.*, 17, 1137, 1975.
21. **Frankenfeld, J. W., Asher, W. J., and N. N. Li**, Biochemical and Biomedical Separations Using Liquid Membranes, presented at the American Chemical Society Symp. Recent Developments in Separation Science, San Francisco, California, August 31, 1976.
22. **Schiffer, D. K., Hochhauser, A., Evans, D. F., and Cussler, E. L.**, *Nature*, 250, 484, August 9, 1974.

Chapter 19

EXTRACTING RESINS

C. Calmon

TABLE OF CONTENTS

I. INTRODUCTION

In the previous two chapters, it has been pointed out that liquid ion exchangers have many advantages over ion exchangers in extraction and elution efficiencies. However, there are disadvantages which reduce their potential use. One major disadvantage is their solubility. To reduce the solubility of the liquid ion exchanger and to enable their use in a column, Small[1] in 1961 and later Clingman and Parrish[2] impregnated various polymeric particles with liquid ion exchangers and showed that these could be used for the extraction and elution of metal ions.

Dow[3] applied macroreticular ion exchangers for containing alkylated amidines for the specific removal of nitrate. Losses of the extractant were observed.

Warshawsky et al.[4] used macroporous polystyrene adsorbents for containing various extractants in a number of hydrometallurgical processes. High loadings could be obtained with commercial liquid ion exchangers in the adsorbents.

Kroebel and Meyer[5] of Farbenfabriken Bayer, Leverkusen, West Germany, have incorporated the liquid in exchangers or extractants into the macroreticular styrene-divinylbenzene-based copolymers during polymerization. Thus, these polymers attain a high porosity with a concomitant improvement in the kinetics, while at the same time, the quantity of extractant that can be incorporated into these polymers is substantially increased. Although the diffusion of the extractant into the aqueous medium is reduced, traces of the extractant in the aqueous medium does create a problem if it is discharged to the environment. Discharge also represents an economic loss. Activated carbon beds can be used for the removal of many of the extractants. To make up for the loss of extractant in the resin particles, reintroduction of the extractant is practiced. Farbenfabriken Bayer terms these extracting resins "Levextrel®." Some of the extractants that may be used are aliphatic amines, phosphoric esters, and both alipatic and aromatic oximes.

II. PROPERTIES OF THE LEVEXTREL RESINS

A. Physical Properties

The beads have a low bulk density (about 630 g/ℓ) and very low specific gravity (about 1.0 g/cm³). The extractant can be 25 to 50% of the bulk weight. Since the extractant is hydrophobic, the resin has a very low water retention (<0.3%).

B. Chemical Properties

The selectivity of these products depends on the extractant and the ions involved. Thus, the variety of products that can be made available is limited only by the number of liquid ion exchangers or liquid extractants available. The resins are an improvement over the liquid/liquid extraction process in that they eliminate the problem of phase separation (i.e. , formation of emulsions).

C. Applications

Applications of these extracting resins to waste effluents are given by Kauczor and Meyer[6] as the following:

1. Removal of traces of toxic metals
2. Removal of radioactive ions
3. Removal of ions in very low concentrations
4. Removal of ions in heavy concentrations but of limited volumes
5. Removal of unwanted traces of ions in concentrated baths
6. Conversion of metal ions from one salt form to another, e.g., $ZnCl_2$ to $ZnSO_4$ through the use of sulfuric acid as the eluting agent
7. Separation of metal ions, thus obtaining salts from wastes in purer form so that they can be more readily recycled

Up to the present time, work has been done with uranium, zinc, plutonium, copper, iron, vanadium, etc. Also, there is the possibility of utilizing such resins with extracts having an affinity for specific organic entities. It must be recognized that the field is new and improved products will be developed especially those having a reduced extractant solubility in the aqueous phase. It is worth watching future developments in this area.

A comparison of ion-exchange resins with the liquid/liquid process and the liquid in a styrene-divinylbenzene polymer is given in Table 1.

TABLE 1

Comparison of Ion-Exchange Resins

Exchange property	Ion-exchange resin	Liquid/liquid process	Liquid in copolymer
Selectivity	+	+ +	+ +
Reaction rate	+	+ +	+
Phase separation	+	0	+
Capacity	+ +	+	+ +
Loss of active groups or substance	None	Measurable	Some
Adsorption of residual concentration	+ +	−	+ +

Note: + +, very good; +, good; 0, moderate; −, poor.

REFERENCES

1. **Small, H.,** Gel-liquid extraction-extraction and separation of metal salts using tributyl phosphate gels, *J. Inorg. Nucl. Chem.,* 18, 232, 1961.
2. **Clingman, A. L. and Parrish, J. R.,** Selective liquid ion exchangers, *J. Appl. Chem.,* 193, 1963.
3. Dow Chemical Co., Nitrate removal from wastewaters, Water Pollution Control Research Series, Report No. 17010-FS, January 1971.
4. **Warshawsky, A. and Patchornik, A.,** Recent developments in metal extraction by solvent impregnated resins, in *Theory and Practice of Ion Exchange,* Society of the Chemical Industry, London, 1976.
5. **Kroebel, R., and Meyer, A.,** Levextrel® a new resin for extract elution, Int. Solvent Extraction Conf. Lyon, France, 1974, 20, 5.
6. **Kauczor, H. W. and Meyer, A.,** Structure and properties of Levextrel resins, in *Theory and Practice of Ion Exchange,* Society of Chemical Industry, London, 1976.

Chapter 20

STARCH-BASED PRODUCTS IN HEAVY METAL REMOVAL

R. E. Wing

TABLE OF CONTENTS

I. INTRODUCTION

Worldwide concern over discharging industrial effluents into waterways arose several years ago. This concern eventually led to strict discharge limits for heavy metals in industrial effluents discharged to receiving waters. Effective waste treatement processes are needed to meet stringent effluent guidelines by 1977 and even more stringent guidelines are proposed for 1983. This urgent need has encouraged researchers to develop new chemical and physical processes for meeting these goals.

In continuing work at the Northern Regional Research Center in Peoria, Illinois to find new uses for starch, several insoluble starch-based products containing xanthate, carboxyl, quaternary ammonium, and tertiary amine groups were prepared and evaluated as potential heavy metal scavengers. Starch, a naturally occurring polymer derived from agricultural crops, is very abundant, annually renewable, and relatively inexpensive. It can be cross-linked with epichlorohydrin and other agents to be made water-insoluble. Chemical modification of the cross-linked starch with various reactive functional monomers yielded products which showed efficacy for the removal of heavy metal ions from waste water.

II. INSOLUBLE STARCH XANTHATE CATION EXCHANGERS

A. Synthesis

A water-insoluble starch xanthate (ISX) offers industry a low-cost product that re-

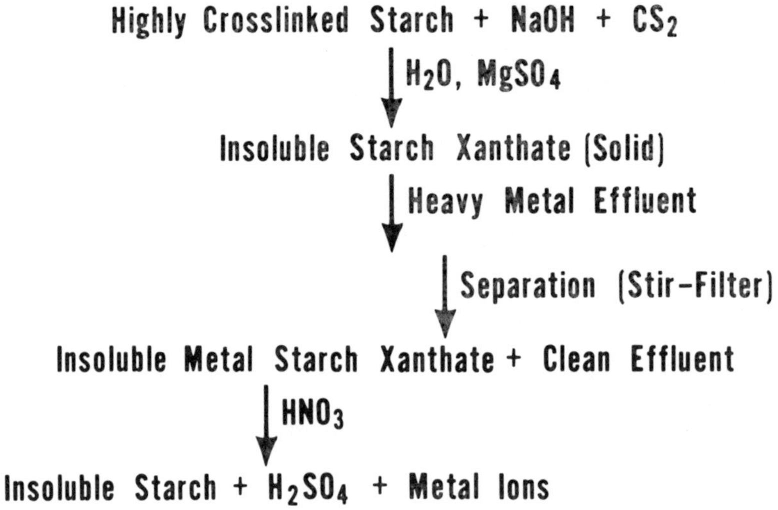

FIGURE 1. Preparation and use of insoluble starch xanthate (ISX).

moves and recovers heavy metals from waste water. Previously, heavy metals had been effectively removed with a starch xanthate-cationic polymer complex[1-3] that allows metal recovery, but almost all cationic polymers are expensive. More traditional removal methods based on inorganic precipitation with lime, alum, or iron salts leave a sludge often placed in landfills. Leaching of the unrecovered metal to underground water and loss of metal resources result. Although precipitation as inorganic salts is inexpensive, recovery of the heavy metals for recycling is difficult.

Starch xanthate made from unmodified corn or wheat starch[4-7] is water soluble. Starch xanthate itself precipitates large amounts of mercury(II), but because almost all other metal xanthate salts are more soluble, removal is not so efficient as when a cationic polymer is included. The efficiencies of metal removal by starch xanthate alone can be greatly increased if the initial starch is cross-linked for water insolubility. Xanthation of this highly cross-linked starch also gives an insoluble product. The addition of magnesium sulfate during the preparation of ISX affords (a) increased filtration or centrifugation rate, (b) increased flash-drying efficiency, (c) increased room temperature stability, and (d) increased settling rates in heavy metal ion removal (see Figure 1).

B. Applications to Heavy Metal Removal

ISX lowers the concentration of practically all heavy metal cations to levels below strict discharge limits.[8-10] The effectiveness of ISX for removing heavy metals from water can be seen in Tables 1 and 2.Table 1 shows the results for solutions (1000 $m\ell$) containing the individual metals of the indicated concentration treated with the indicated amount of ISX (capacity = 1.56 meq/g) at pH 3.7. Solutions were stirred for 5 to 60 min to a final pH of 8.9. After filtration, the residual metals were determined by a Varian Techtron® AA 120. The theoretical weight of ISX for a divalent metal is 0.64 g. The value denoted (<) was below the detection limits of the instrument.

Table 2 shows the results for a solution (1000 $m\ell$) containing a mixture of heavy metals of the indicated concentrations at pH 3.5 treated with ISX (capacity = 1.56 meq/g, 0.32 g) to a final pH of 8.9. After filtration, the residual metals were also determined by a Varian Techtron® AA 120. Values denoted (<) were below the detection limits of the instrument. The data show that 12 heavy metals are removed from

TABLE 1

Removal of Heavy Metal Cations from Water with Insoluble Starch Xanthate (ISX)

Metal	Initial concentration (mg/ℓ)	I S X (g)	Residual concentration (mg/ℓ)	Illinois discharge limit (mg/ℓ)
Ag^{+}	53.94	0.32	0.016	0.005
Au^{+3}	30.00	0.50	<0.010	—
Cd^{+2}	56.20	0.64	0.012	0.050
Co^{+2}	29.48	0.64	0.090	—
Cr^{+3}	26.00	0.64	0.024	1.0
Cu^{+2}	31.77	0.32	0.008	0.020
Fe^{+2}	27.92	0.32	0.015	1.0
Hg^{+2}	100.00	0.64	0.001	0.0005
Mn^{+2}	27.47	0.64	0.015	1.0
Ni^{+2}	29.35	0.64	0.160	1.0
Pb^{+2}	103.60	0.64	0.035	0.100
Zn^{+2}	32.69	0.32	0.294	1.0

TABLE 2

Removal of Metals from a Dilute Solution with Insoluble Starch Xanthate (ISX)

Metal	Initial concentration (mg/ℓ)	Residual concentration (mg/ℓ)
Cd^{+2}	5.62	0.001
Co^{+2}	2.95	<0.010
Cr^{+3}	2.60	0.026
Cu^{+2}	3.18	<0.005
Fe^{+2}	2.79	0.001
Hg^{+2}	10.00	0.0007
Mn^{+2}	2.75	0.010
Ni^{+2}	2.93	<0.050
Pb^{+2}	10.36	<0.031
Zn^{+2}	3.27	0.007

concentrated solutions and dilute solutions, usually to concentrations below present discharge limits.

Table 3 shows the results of heavy metals removal from some actual industrial samples. Solutions (1000 mℓ) of industrial effluents were treated with the indicated amounts of ISX (capacity = 1.1 meq/g). After stirring 5 min, the solutions were filtered and the residual metal concentrations determined by atomic absorption.

From these data, ISX appears to be effective in removing metal ions at different concentration levels. If initial metal concentrations in an industrial effluent exceed 100 mg/l, it may not be economical to use ISX, and removal by chemical precipitation or another process could be used. The ISX could then be used in a secondary treatment to further lower metal concentrations below discharge limits. From studies on ISX to date, the following statements can be made:

1. Its average capacity is 1.1 to 1.5 meq of metal per gram of ISX.
2. It is effective over a pH range of 3 to 11, with maximum effectiveness above pH 7.
3. Salt concentrations of up to 10% have little influence on the effectiveness of ISX.

TABLE 3

Industrial Effluents Treated with Insoluble Starch Xanthate (ISX)

Industrial effluent sample	Initial pH	ISX (g)	Final pH	Cd^{+2} (mg/ℓ)	Cr^{+3} (mg/ℓ)	Cu^{+2} (mg/ℓ)	Fe^{+2} (mg/ℓ)	Pb^{+2} (mg/ℓ)
A								
Initial	5.3	—	—	0.150	19.4	0.105	—	—
Treated	—	0.416	7.3	0.004	0.013	0.009	—	—
B								
Initial	5.3	—	—	0.087	250.0	0.150	3.04	0.55
Treated	—	1.60	6.7	0.005	0.216	0.016	0.001	0.001
C								
Initial	4.5	—	—	4.197	20.0	141.0	—	13.11
Treated	—	2.45	7.1	0.015	0.017	0.10	—	0.166

4. ISX instantaneously removes heavy metal ions.
5. Treatment is applicable to batch-type or continuous-flow, or filter precoat systems.
6. ISX metal sludge settles rapidly in the quiescent state and dewaters to 50 to 90% solids content after filtration or centrifugation.
7. Metals can be recovered from the ISX metal sludge by nitric acid treatment or incineration.
8. Preliminary cost-to-make estimate for ISX is $0.30/lb.

III. STARCH-BASED ANION EXCHANGERS

A. Synthesis of Tertiary Amine- and Quaternary Ammonium-Cross-linked Starch

Commercial anion-exchange resins containing tertiary amine and quaternary ammonium groups are very useful in industrial wastewater treatment to remove heavy metal anions.[11,12] The initial cost to install these exchange resin systems has limited the industrial application of this antipollution technique. The main reason for the high cost is that these resins are petrochemically based and because of the present petrochemical shortages, these products will likely increase in cost and may even be limited in supply. A possible solution to this problem might be to use a naturally occurring, annually renewable, low-cost polymer such as starch.

The preparation and use of water-soluble cationic starches is well known.[13] Cationic functionalities attached to cellulose or cross-linked dextrans[14-18] are water insoluble and in a highly purified form have been useful in chromatography. These products usually only have moderate capacity (<1.0 meq/g) and are expensive.

Very little work has been published using cross-linked starch (CLS)[19] as a water-insoluble matrix for cationic monomers.[20-23] Insoluble starch products containing tertiary amine (TA-CLS) and quaternary ammonium (QA-CLS) ethers of different degrees of substitution and capacities in the range of 2.12 to 2.75 meq/g were prepared and evaluated as heavy metal anion scavengers.

B. Applications to Heavy Metal Removal

These starch products (TA-, QA-CLS) can be added directly to the polluted wastewater stream containing heavy metal anions. Examples which show the effectiveness of the starch-based products include the removal of chromate, dichromate, ferricyanide, ferrocyanide, permanganate, and molybdate ions. Generally, the starch-based products

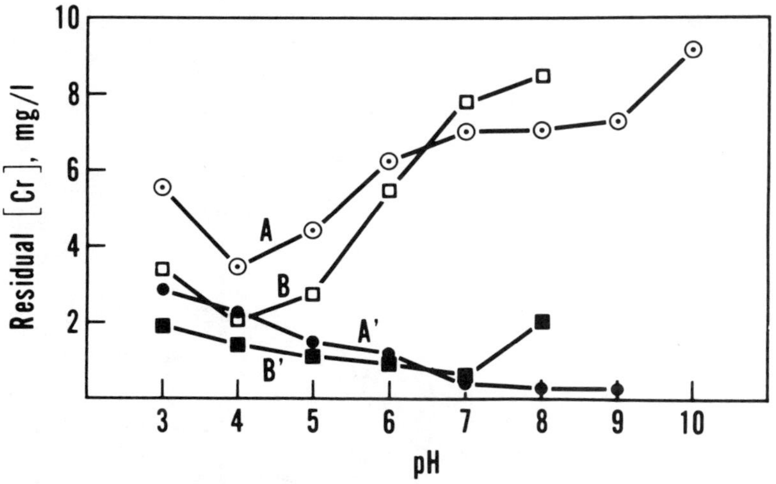

FIGURE 2. Chromate removal with QA-CLS and TA-CLS vs. pH. Chromate ($CrO_4^{=}$) solutions (250mℓ, 26 mg/ℓ as Cr^{+6}) were treated with QA-CLS (capacity = 2.13 meq/g; A, 0.10 g; A, 0.20 g) and TA-CLS (capacity = 2.75 meq/ℓ; B, 0.10 g; B, 0.20 g) at the indicated pH.

are regenerable like the petroleum-based ion exchange resins. Heavy metal anions, such as $HgCl^{-2}_4$ and $AuCl_4^-$, should also be removed by these starch-based products.

1. Chromate and Dichromate

The use of chromates ($CrO_4^{=}$) and dichromates ($Cr_{27}^{=}$) in plating and as corrosion control agents in cooling water systems is quite extensive. Chromium(VI) is very toxic and must be removed from industrial wastewaters before discharge. Chromium(VI) levels allowed in waste effluents which are discharged to natural waterways have been established at a maximum concentration of 0.05 mgℓ. Rinse waters of plating operations may vary in composition, but they usually contain 20 to 100 mgℓ Cr^{+6} and have a pH of 2 to 3. These rinses are treated for three reasons: pollution control, water reuse, and Cr^{+6} recovery.

Treatment for Cr^{+6} usually involves chemical reduction to Cr^{+3} with sulfur dioxide, sodium bisulfite, sodium metabisulfite, or ferrous salts[24-28] at pH 2 to 3. The Cr^{+3} is then precipitated at alkaline pHs as the hydroxide with caustic or lime. Nonreductive treatments of Cr^{+6} include anion exchange[29-34] and activated carbon.[35] Since anion-exchange resins are effective in the removal of Cr^{+6}, the starch-based cationic products were evaluated.

The TA-CLS and the QA-CLS were evaluated to determine their optimum pH range and capacity for Cr^{+6} as chromate and dichromate. Figures 2 to 4 show these results. Figure 2 shows the optimum pH for removal of chromate at higher concentrations of Cr^{+6} to be around 4; however, as the Cr^{+6} concentration decreases to 3 to 4 mg/ℓ, a pH near 7 is optimum to reach levels near the discharge limit of 0.05 mg/ℓ. These studies were run with less starch product than theoretically required. The capacity of QA-CLS and TA-CLS for chromate is shown in Figures 3 and 4. These figures again show good Cr^{+6} removal at pH 4 for the higher Cr^{+6} concentrations, with pH 7 giving the lowest residual concentrations. These figures also show that over 90% of the Cr^{+6} is removed when the theoretical amount of the starch product is used. The pH and capacities of the products for dichromate are very similar to those for chromate.

These starch-based products are effective in both the chloride and hydroxide forms,

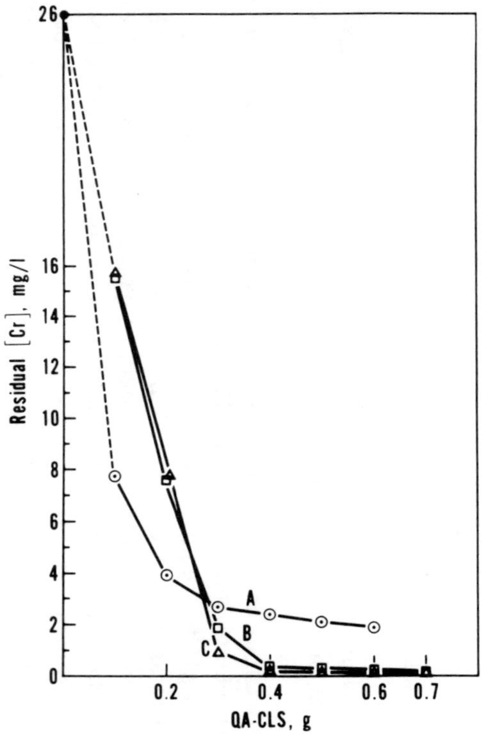

FIGURE 3. Chromate removal with QA-CLS at pH 4, 7, and 9. Chromate ($CrO_4^{=}$) solutions (500 mℓ, 26 mg/ℓ as Cr^{+6}) at pH 4 (A), 7 (B), and 9 (C) were treated with increasing amounts of QA-CLS (capacity = 2.13 meq/g). Aliquots (10mℓ) of the supernatant were removed for chromium analysis 10 min after each addition. Theoretical weight required is 0.23 g.

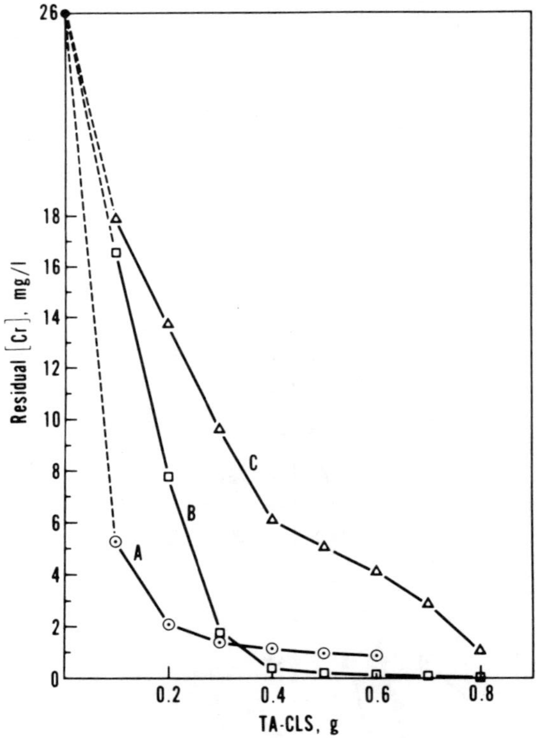

FIGURE 4. Chromate removal with TA-CLS at pH 4, 7, and 9. Chromate ($CrO_4^{=}$) solutions (500 mℓ, 26mg/ℓ as Cr^{+6}) at pH 4 (A), 7 (B), and 9 (C) were treated with increasing amounts of TA-CLS (capacity = 2.75 meq/g). Aliquots (10 mℓ of the supernatant were removed for chromium analysis 10 min after each addition. Theoretical weight required is 0.18 g.

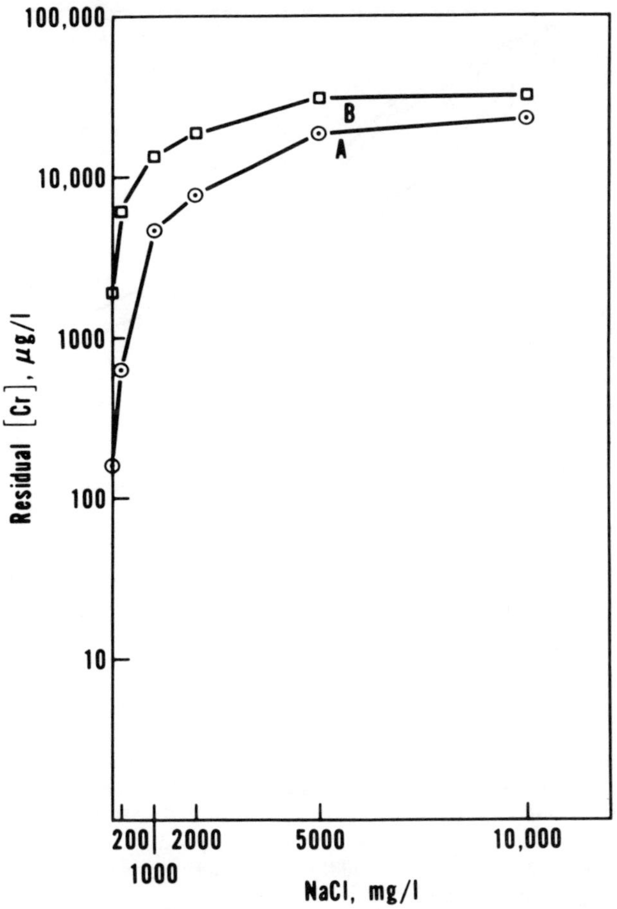

FIGURE 5. Chromate removal in the presence of salt (NaCl) with QA-CLS and TA-CLS. A. Chromate ($Cr_2O_4^=$) solutions (50 mℓ, 26 mg/ℓ as Cr^{+6}) containing various amounts of sodium chloride were treated at pH 6 with QA-CLS (0.1 g, 2.13 meq/g). B. Dichromate ($Cr_2O_7^=$) solutions (50 mℓ, 49.7 mg/ℓ as Cr^{+6}) containing various amounts of sodium chloride were treated at pH 4 with TA-CLS (0.05 g, 2.75 meq/g).

and usually no pretreatment would be necessary as with typical ion-exchange columns. To reach discharge limit levels, a slight excess of product is necessary. The oxidative nature of chromate and dichromate ions does not appear to affect these starch-based products. The chromate or dichromate is released from these products by washing with sodium hydroxide (1 *N*), hydrochloric acid (2%), or sodium chloride (5%). The regenerated product is still effective in Cr^{+6} removal. Since sodium chloride is effective in removing Cr^{+6} from the resin, these starch-based products would not be expected to tolerate high salt concentrations in the treatment of industrial effluents. The salt effect is shown in Figure 5. If sulfate ions are present, the removal will be slightly lower. If salt concentrations are too high, the effluent would have to be diluted with other process waters to lower the salt concentration in order to have effective Cr^{+6} removal with these starch-based cationic polymers. The products could be pelletized for use in pressurized columns having acceptable flowrates to overcome this salt effect to some extent.[36] Figure 6 shows the use of the starch-based products for a chromic acid industrial

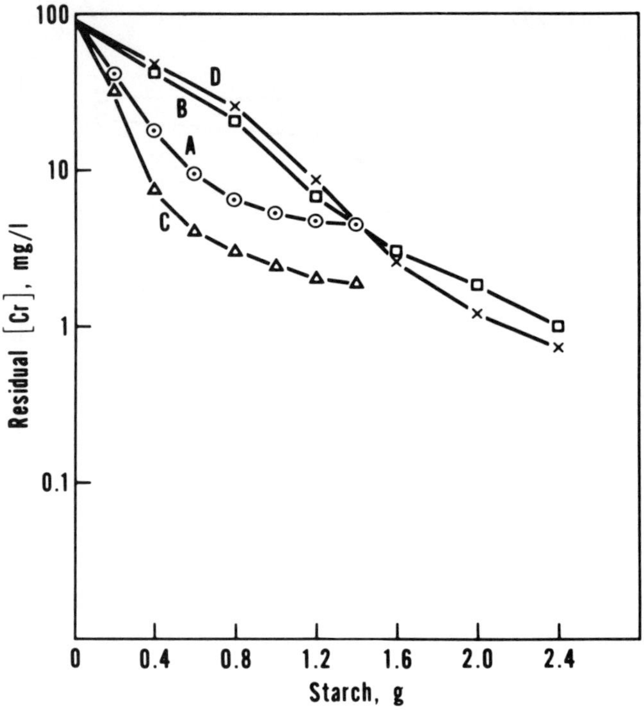

FIGURE 6. Chromate removal from an industrial chromic acid rinse. Solutions (500 mℓ, 92 mg/ℓ Cr) at pH 4 and 7 were treated with increasing amounts of QA-CLS (capacity = 2.13 meq/g; A, pH 4; B, pH 7) and TA-CLS (capacity = 2.75 meq/g; C, pH 4; D, pH 7). Aliquots (10 mℓ) of the supernatant were removed for chromium analysis 10 min after each addition. Theoretical weight required is 0.83 g and 0.64 g, respectively.

effluent. The results previously described are also evident in these curves. Small amounts of Cr^{+3} (1 mg/ℓ) are present, which prevent total chromium removal below 1 mg/ℓ. This particular effluent contains several other heavy metal cations which were not removed by these products.

2. Iron, Cyanide Complexes

Aqueous cyanide pollution in industrial wastes causes widespread problems. Several methods of cyanide removal are used today. These include alkaline chlorination, electrolytic oxidation, ozonation, ion-exchange, Kastone® (H_2O_2-formaldehyde), and activated carbon.[37-42] Recently, an ion-exchange process was developed to convert free cyanides into the very stable ferrocyanide complex and then to remove the complex on a weakly basic ion-exchange resin.[43,44]

The tertiary amine- and quaternary ammonium-cross-linked starches have been evaluated for ferricyanide ($Fe(CN_6)^{-3}$) and ferrocyanide ($Fe(CN_6)^{-4}$) removal. Figures 7 to 12 show the most effective pH range, the capacity of the products, and their effectiveness in the presence of 1 to 4% salt (NaCl). Figures 7 and 8 show that a pH range of 4 to 5 is optimum. The products evidence a strong binding capacity, as shown in Figures 9 and 10, since no detectable amount of complex was found in the treated solutions. The presence of salt slightly limits $Fe(CN)_6^{-3}$ and $Fe(CN)_6^{-4}$ removal, as shown in Figures 11 and 12. A 1% NaOH solution treatment effectively regenerates the products.

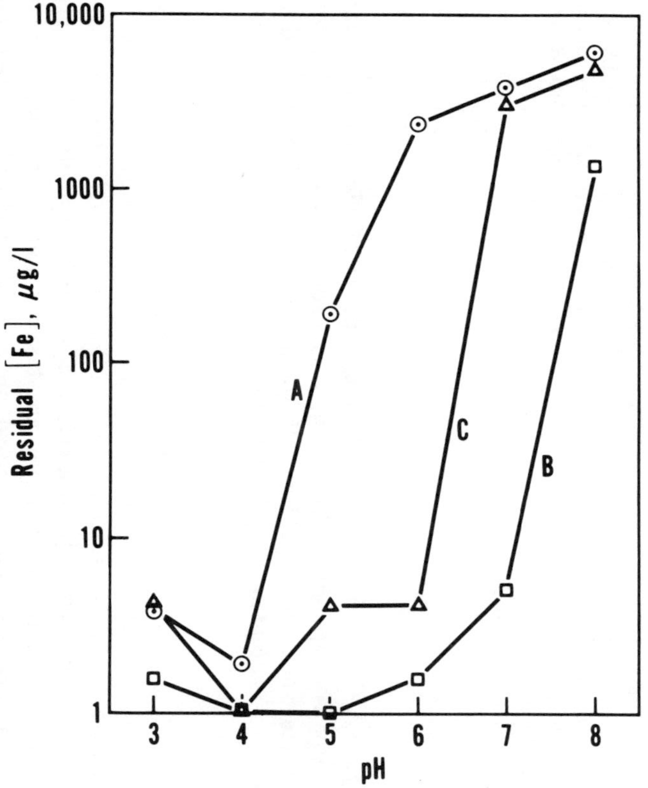

FIGURE 7. Ferricyanide and ferrocyanide removal with TA-CLS vs. pH. Solutions (50 mℓ were treated with TA-CLS (capacity = 2.75 meq/g) at the indicated pH. A. Ferricyanide concentration (25.5 mg/ℓ as Fe) with 0.035 g TA-CLS. B. Ferricyanide concentration (25.5 mg/ℓ as Fe) with 0.05 g TA-CLS. C. Ferrocyanide concentration (25.9 mgℓ as Fe) with 0.055 g TA-CLS.

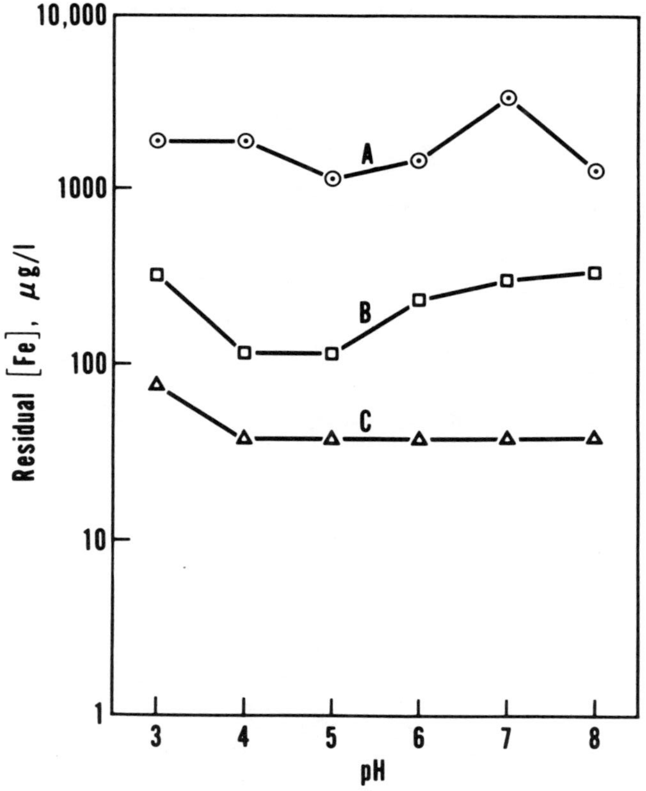

FIGURE 8. Ferricyanide and ferrocyanide removal with QA-CLS vs. pH. Solutions (50 mℓ) were treated with QA-CLS (capacity = 2.13 meq/g) at the indicated pHs. A. Ferricyanide concentration (25.5 mg/ℓ as Fe) with 0.04 g QA-CLS. B. Ferricyanide concentration (25.5 mg/ℓ as Fe) with 0.045 g QA-CLS. C. Ferrocyanide concentration (25.9 mg/ℓ as Fe) with 0.065 g QA-CLS.

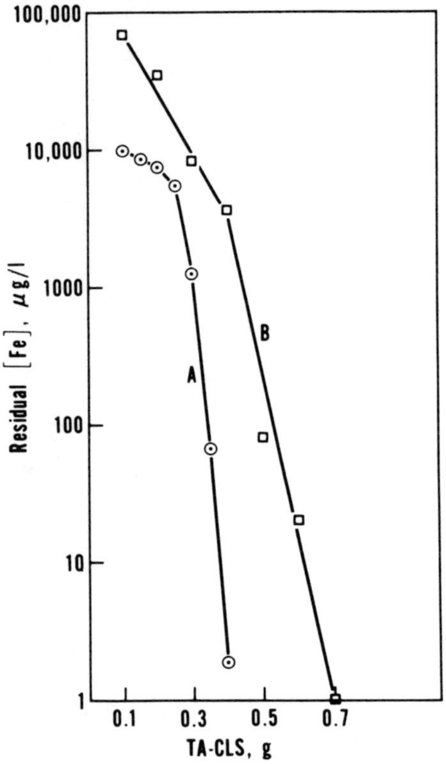

FIGURE 9. Ferricyanide and ferrocyanide removal with TA-CLS. Solutions (500 mℓ) at pH 5 were treated with increasing amounts of TA-CLS (capacity = 2.75 meq/g). Aliquots (10 mℓ) of the supernatant were removed for iron analysis after each addition. A. Ferricyanide concentration (25.5 mg/ℓ as Fe, initial); theoretical weight of TA-CLS required, 0.25 g. B. Ferrocyanide concentration (25.9 mg/ℓ as Fe, initial); theoretical weight of TA-CLS required, 0.34 g.

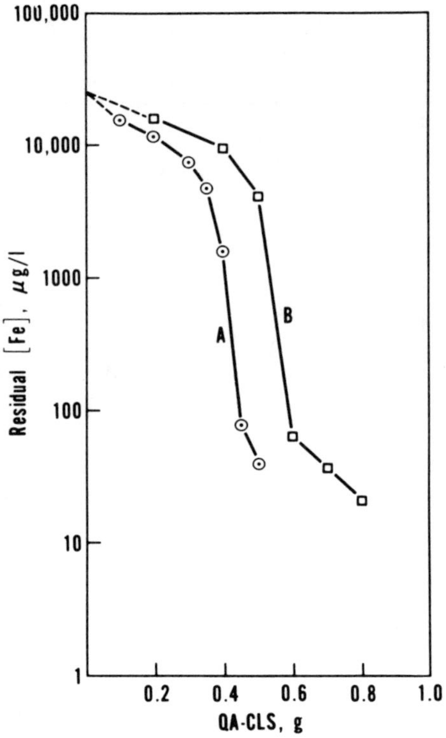

FIGURE 10. Ferricyanide and ferrocyanide removal with QA-CLS. Solutions (500 mℓ) at pH 5 were treated with increasing amounts of QA-CLS (capacity = 2.13 meq/g). Aliquots (10 mℓ) of the supernatant were removed for iron analysis 10 min after each addition. A. Ferricyanide concentration (25.5 mg/ℓ as Fe, initial); theoretical weight of QA-CLS required, 0.32 g. B. Ferrocyanide concentration (25.9 mg/ℓ as Fe, initial); theoretical weight of QA-CLS required, 0.44 g.

3. Molybdate

The QA-CLS and TA-CLS were evaluated in the removal of Mo^{+6} as molybdate. Figures 13 and 14 show corresponding pH and capacity data for this anion. Optimum molybdate ($MoO_4^{=}$) removal occurs at pH 4, with theoretical amounts of the products removing $MoO_4^{=}$ close to its nondetectable limit (0.1 mgℓ).

4. Permanganate

QA-CLS and TA-CLS are very effective in lowering manganese(VII) levels from solutions of permanganate. Figures 15 and 16 show that optimum removal occurs at pH 7 and that the QA-CLS is slightly more effective than the TA-CLS at theoretical capacity. Even though these products lower the Mn(VII) level well below established discharge limits, they turn brown-purple due to oxidation and become nonregenerable. Since the product is very efficient in permanganate removal, an excess of product is not required.

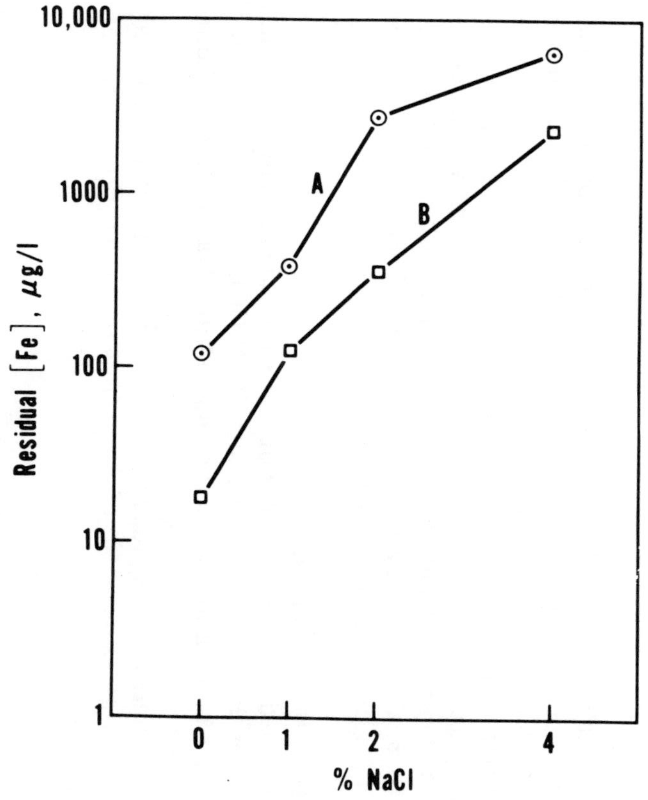

FIGURE 11. Ferrocyanide removal in presence of NaCl with QA-CLS and TA-CLS. Solutions (50 mℓ) of ferrocyanide (25.9 mg/ℓ as Fe) containing various amounts of sodium chloride were treated at pH 4. A. QA-CLS (capacity = 2.12 meq/g, 0.065 g). B. TA-CLS (capacity = 2.75 meq/g. 0.055 g).

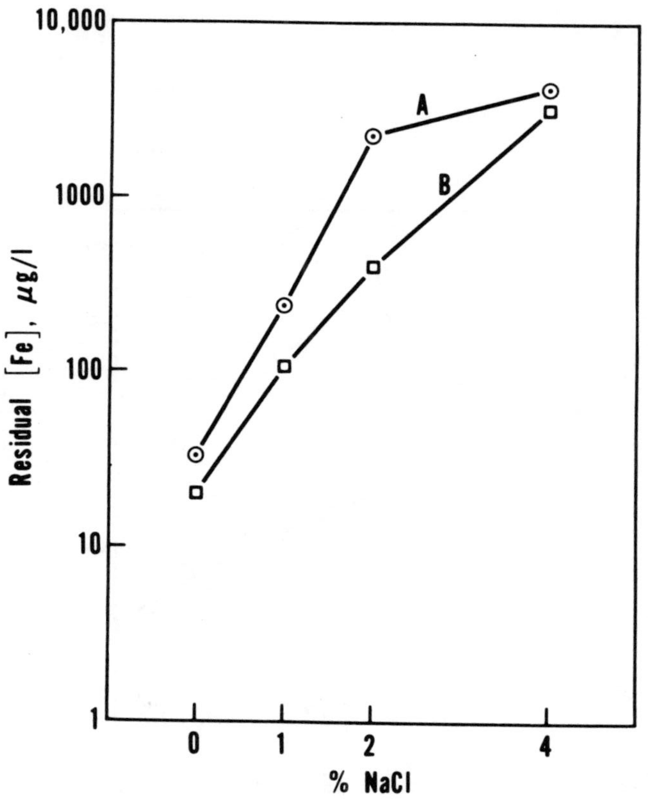

FIGURE 12. Ferricyanide removal in presence of NaCl with QA-CLS and TA-CLS. Solutions (50 mℓ) of ferricyanide (25.5 mg/ℓ as Fe) containing various amounts of sodium chloride were treated at pH 4. A. QA-CLS (capacity = 2.12 meq/g. 0.05 g). B. TA-CLS (capacity = 2.75 meq/g, 0.04 g).

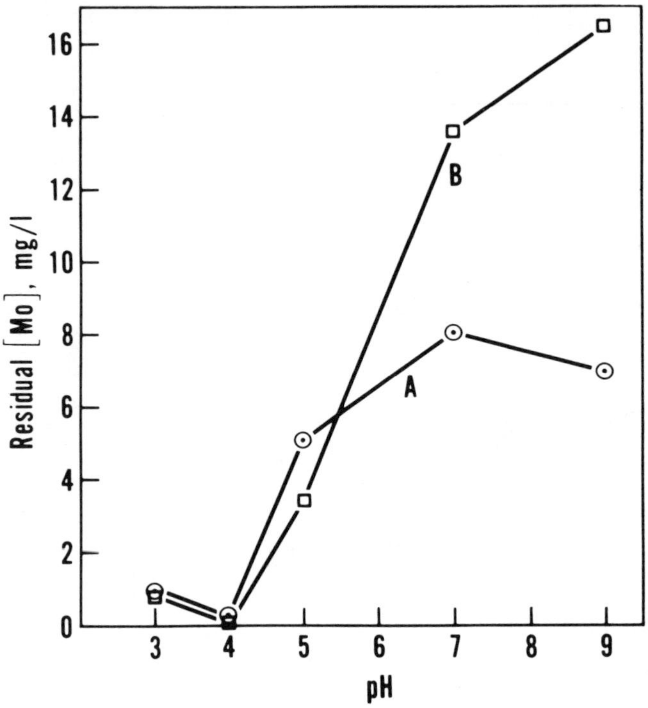

FIGURE 13. Molybdate removal with QA-CLS and TA-CLS vs. pH. Molybdate ($MoO_4^{=}$) solutions (50 mℓ, 47.97 mg/ℓ as Mo^{+6}) were treated at the inidicated pH. A. QA-CLS (capacity = 2.13 meq/g, 0.03 g). B. TA-CLS (capacity = 2.75 meq/g, 0.03 g).

IV. CARBOXYL CROSS-LINKED STARCH CATION EXCHANGERS

A. Synthesis

Two types of water-insoluble carboxyl-containing products were prepared and evaluated in heavy metal cation removal. Carboxymethyl cross-linked starch (CCLS) was prepared by reaction of chloroacetic acid, base, and cross-linked starch to yield a product containing monomeric CH_2COO^- groups on the starch backbone (capacity = 1.43 meq/g). The other product is a grafted cross-linked starch (GCLS) containing numerous carboxyl groups in long side chains (capacity = 2.92 meq/g). Some selected examples show their potential in removing heavy metal cations. These products are totally regenerable with 2% HCl, allowing their reuse.

B. Applications to Heavy Metal Removal

The GCLS was evaluated over the pH range of 3 to 8 for cadmium removal; a pH 7 gave optimum removal. Figure 17 shows the capacity of GCLS and CCLS for cadmium and how effective these products are for cadmium removal. Figure 18 shows similar results for copper.

The mention of firm names or trade products does not imply that they are endorsed or recommended by the U.S. Department of Agriculture over other firms or similar products not mentioned.

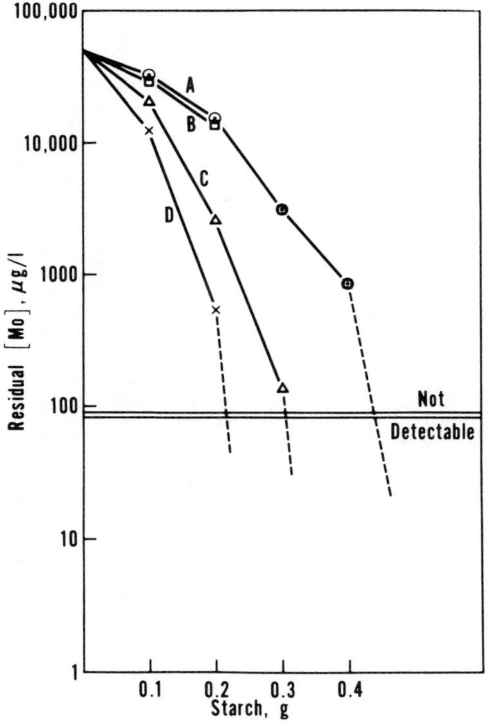

FIGURE 14. Molybdate removal with QA-CLS and TA-CLS at pH 4 and 7. Molybdate ($MoO_4^{=}$) solutions (500 mℓ, 47.97 mg/ℓ as Mo^{+6} initial concentration) were treated with increasing amounts of TA-CLS. (capacity = 2.75 meq/g; A. pH 7; D. pH 4; theoretical weight required, 0.11 g) and QA-CLS (capacity = 2.13 meq/g; B. pH 7; C. pH 4; theoretical weight required, 0.14 g). Aliquots (10 mℓ) of the supernatant were removed after 10 min for molybdenum analysis after each addition.

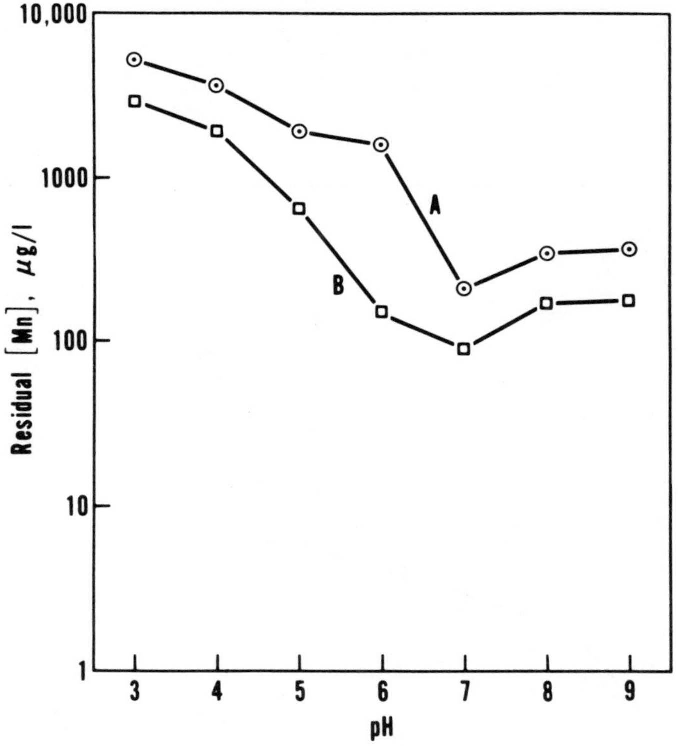

FIGURE 15. Permanganate removal with QA-CLS and TA-CLS vs. pH. Permanganate (MnO_4^-) solutions (50 mℓ, 28.6 mg/ℓ as Mn^{+7}) were treated at the indicated pH. A. QA-CLS (capacity = 2.13 meq/g, 0.05 g). B. TA-CLS (capacity = 2.75 meq/g, 0.05 g).

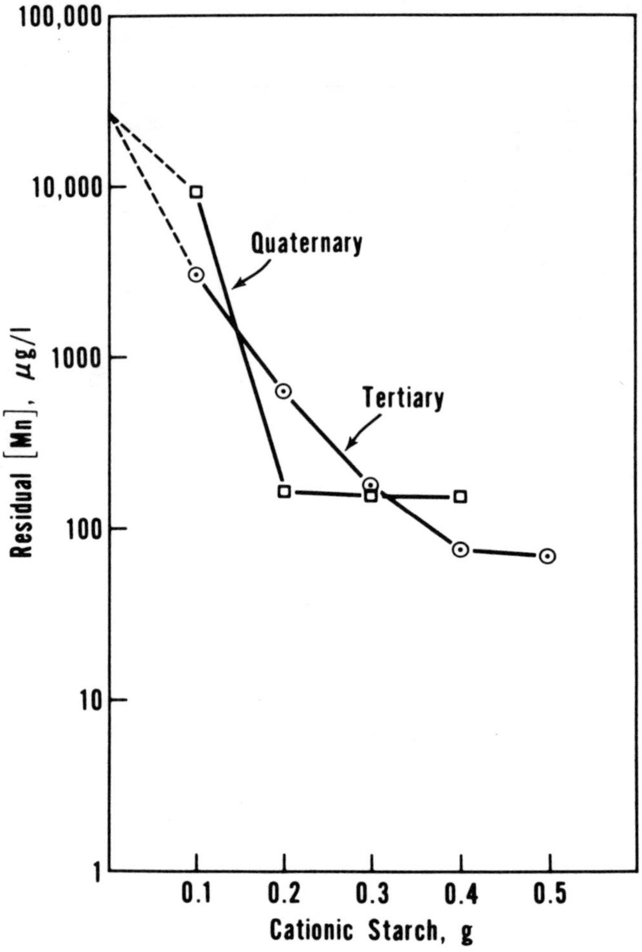

FIGURE 16. Manganese (VII) removal with QA-CLS and TA-CLS. Permanganate (MnO_4^-) solutions (500 mℓ, 28.6 mg/ℓ as Mn^{+7} initial concentration) were treated with increasing amounts of QA-CLS (capacity = 2.13 meq/g; A. theoretical weight required, 0.23 g) and TA-CLS (capacity = 2.75 meq/g; B. theoretical weight required, 0.18 g). Aliquots (10 mℓ) of the supernatant were removed after 10 min for manganese analysis after each addition.

REFERENCES

1. **Swanson, C. L., Wing, R. E., Doane, W. M., and Russell, C. R.,** Mercury removal from waste water with starch xanthate-cationic polymer complex, *Environ. Sci. Technol.*, 7 (7), 614, 1973.
2. **Wing, R. E., Swanson, C. L., Doane, W. M., and Russell, C. R.,** Heavy metal removal with starch xanthate-cationic polymer complex, *Water Pollut. Contr. Fed. J.*, 46 (8), 2043, 1974.
3. **Swanson, C. L., Wing, R. E., and Doane, W. M.,** Removal of heavy metal ions from waste water, U.S. Patent 3,947,354, March 30, 1976.
4. **Doane, W. M., Russell, C. R., and Rist, C. E.,** Uniformity of starch xanthates prepared by a rapid, continuous process, *Staerke*, 3, 77, 1965.
5. **Lancaster, E. B., Black, L. T., Conway, H. F., and Griffin, E. L., Jr.,** Xanthation of starch in low-concentration pastes, *Ind. Eng. Chem., Prod. Res. Dev.*, 5 (4), 354, 1966.

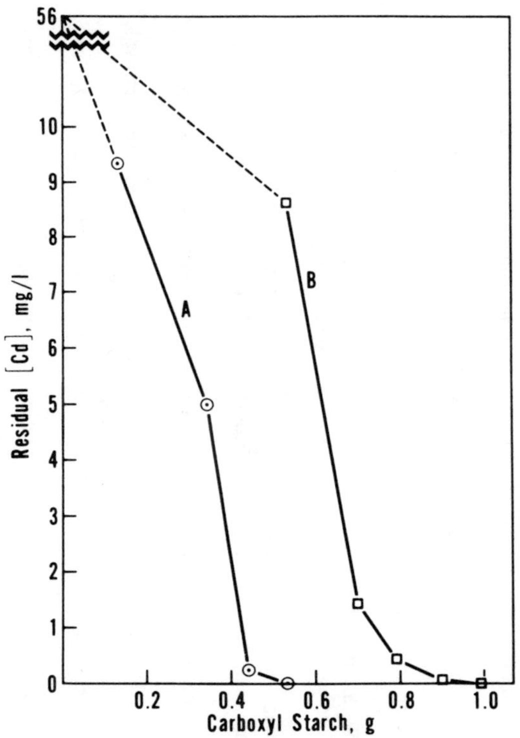

FIGURE 17. Cadmium removal with carboxylated cross-linked starches at pH 7. Cadmium (CD^{+2}) solutions (1000 mℓ, 56.2 mg/ℓ as Cd^{+2}, initial concentration) were treated at pH 7 with increasing amounts of grafted cross-linked starch (GCLS) (capacity = 2.92 meq/g; A. theoretical weight required, 0.34 g) and carboxylated cross-linked starch (CCLS) (capacity = 1.43 meq/g; B. theoretical weight required, 0.69 g). Aliquots (10 mℓ) of the supernatant were removed for cadmium analysis after each addition.

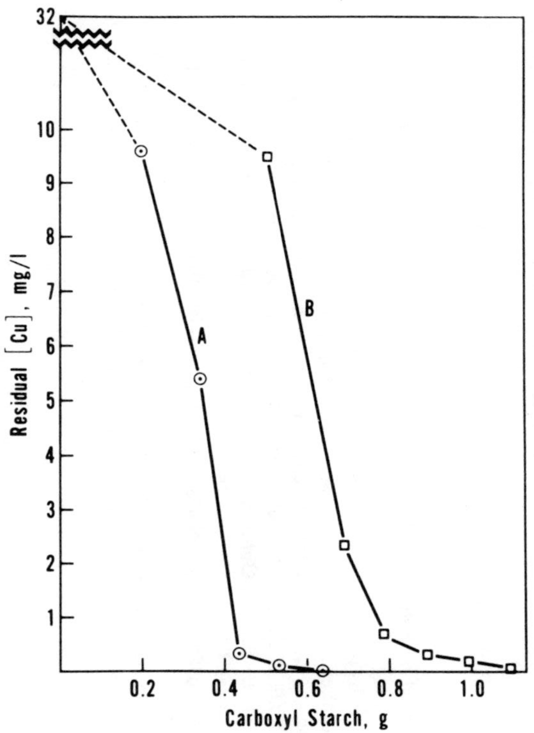

FIGURE 18. Copper removal with carboxylated cross-linked starches at pH 5. Copper (Cu^{+2}) solutions (1000 mℓ, 31.8 mg/ℓ Cu^{+2}, initial concentration) were treated at pH 5 with increasing amounts of GCLS (capacity = 2.92 meq/g; A. theoretical weight required, 0.34 g) and CCLS (capacity = 1.43 meq/g; B. theoretical weight required, 0.69 g). Aliquots (10mℓ) of the supernatant were removed for copper analysis after each addition.

6. **Russell, C. R., Buchanan, R. A., Rist, C. E., Hofreiter, B. T., and Ernst, A. J.,** Cereal pulps. 1. Preparation and application of cross-linked cereal xanthates in paper products, *Tappi,* 45 (7), 557, 1962.
7. **Swanson, C. L., Naffziger, T. R., Russell, C. R., Hofreiter, B. T., and Rist, C. E.,** Xanthation of starch by a continuous process, *Ind. Eng. Chem., Prod. Res. Dev.,* 3 (1), 22, 1964.
8. **Wing, R. E., Doane, W. M., and Russell, C. R.,** Insoluble starch xanthate: use in heavy metal removal, *J. Appl. Polym. Sci.,* 19 (3), 847, 1975.
9. **Wing, R. E.,** Corn starch compound recovers metal from water, *Ind. Wastes* (Chicago), 21 (1), 26, 1975.
10. **Wing, R. E. and Doane, W. M.,** Removal of heavy metal ions from aqueous solutions with insoluble crosslinked starch xanthate, U.S. Patent 3,979,286, September 7, 1976; 4,051,316, September 27, 1977.
11. **Higgins, T. R.,** Ion exchange: its present and future use, *Environ. Sci. Technol.,* 7, 1110, 1973.
12. **Kunin, R.,** Ion exchange for the metal products finisher, *Prod. Finish.* (London), 33, 71, 1969.
13. **Paschall, E. F.,** Production and uses of cationic starches, in *Starch: Chemistry and Technology,* Vol. 2, Whistler, R. L. and Paschall, E. F., Eds. Academic Press, New York, 1967, 403.
14. **Benerito, R. R., Woodward, B. B., and Guthrie, J. D.,** Preparation and properties of quaternary cellulose anion exchangers, *Anal. Chem.,* 37, 1693, 1965.
15. **Kuniak, L.,** Crosslinked diethylaminoethyl cellulose, Czechoslovakian Patent 152, 781, April 15, 1974.
16. Selectacel Data Sheets, Brown Company, Berlin-Gorham Division, Berlin, N. H., December, 1974.
17. **W. & R. Balston, Ltd.,** Improvements in or relating to materials containing ion exchange groups, British Patent 1,001,481, August 18, 1965.
18. **Aktiebolaget Pharmacia,** Improvements in or relating to ion-exchange materials, British Patent 1,013,585, December 15, 1965.
19. **Hullinger, C. H.,** Production and use of crosslinked starch in *Starch: Chemistry and Technology,* Vol. 2, Whistler, R. L. and Paschall, E. F., Eds., Academic Press, New York, 1967, 445.
20. **Kuniak, L. and Marchessault, R.,** Cross-linked ion exchange starches, Canadian Patent 960, 652, January 7, 1975.
21. **Aktiebolaget Pharmacia,** Preparation of substituted high molecular weight products, British Patent 936,039, September 4, 1963.
22. **Bullock, A. L. and Guthrie, J. D.,** Cation exchange starches which retain their original granular form and process for making same, U.S. Patent 3,065,222, November 20, 1962.
23. **Bullock, A. L. and Guthrie, J. D.,** Ion exchange starches which retain their original granular form and process for making same, U.S. Patent 2,992,215, July 11, 1961.
24. **Leimkuehler, K. C.,** How NASA removes hexavalent chrome, *Ind. Wastes,* (Chicago), 21 (5), 38, 1975.
25. **Hill, G. B.,** Complete removal of chromic acid waste with the aid of instrumentation, *Plating,* (East Orange, N.J.), 56, 172, 1969.
26. **Zievers, J. F. and Novotny, C. J.,** Recovery of chromium plating rinse water, U.S. Patent 3, 681, 2110, August 1, 1972.
27. **Landy, J. A.,** Chromate removal at a Saudi Arabian fertilizer complex, *Water Pollut. Control Fed. J.,* 43, 2242, 1971.
28. **Onstott, E. I., Gregory, W. S., and Thode, E. F.,** Removal of chromate from cooling tower blowdown by reaction with electro-chemically generated ferrous hydroxide, *Environ. Sci. Technol.,* 7, 333, 1973.
29. **Rothstein, S.,** Five years of ion exchange, *Plating,* (East Orange, N.J.), 45, 835, 1958.
30. **Arden, T. V. and Giddings, M.,** Anion exchange in chromate solution, *J. Appl. Chem.,* 11, 229, 1961.
31. **Oberhofer, A. W.,** Method for selectively removing chromates, U.S. Patent 3,414,510, December 3, 1968; 3,223,620, December 14, 1965.
32. **Saraceno, A. J., Walters, R. H., Jones, D. B., and Wiehle, W. E.,** Process for selective removal and recovery of chromates from water, U.S. Patent 3,664,950, May 23, 1972.
33. **Sloan, L. and Nitti, N. J.,** Method and apparatus for fluid treatment, U.S. Patent 3,306,859, February 28, 1967.
34. **Zievers, J. F., Riley, C. W., and Crain, R. W.,** Metal ion recovery system, U. S. Patent 3,658,470, April 25, 1972.
35. **Linstedt, K. D., Houck, C. P., and O'Connor, J. T.,** Trace element removals in advance wastewater treatment processes, *Water Pollut. Control Fed. J.,* 43, 1507, 1971.
36. **Zievers, J. F.,** Industrial Filter and Pump Manufacturing Corp., Cicero, Ill., personal communication.

37. **Zievers, J. F., Crain, R. W., and Barclay, F. G.**, Metal finishing wastes: methods of disposal, *Plating*, (East Orange, N.J.), 57, 56, 1970.
38. **Beevers, M.**, Chlorine and sulfur dioxide in the treatment of cyanide and chromium wastes, *Met. Finish. J.*, 18, 232, 1972.
39. **Sastry, V. V. R and Mahapatra, S.**, Utilization and treatment of waste water in metal finishing industries, *Indian Chem. J.*, 12 (9), 27, 1974.
40. **Sulfaro, A. N. and Stein, A. H.**, Apparatus and process for treating toxic waste materials, U.S. Patent 3,715,308, February 6, 1973.
41. **Mehrkam, Q. D.**, A process for disposal of salt bath wastes, *Met. Prog.*, 108(4), 103, 1975.
42. **Anon.**, New process detoxifies cyanide wastes, *Environ. Sci. Technol.*, 5, 496, 1971.
43. **Fries, W.**, Selective separation of cyanide from waste streams by adsorption process, U.S. Patent 3,788,983, January 29, 1975.
44. **Avery, N. L. and Fries, W.**, Selective removal of cyanide from industrial waste effluents with ion-exchange resins, *Ind. Eng. Chem., Prod. Res. Dev.*, 14, 102, 1975.

Chapter 21

ACTIVATED CARBON FOR THE REMOVAL OF TRACE METALS FROM AQUEOUS SOLUTIONS

P. C. Singer

TABLE OF CONTENTS

I. NATURE OF ACTIVATED CARBON[1,2]

Activated carbons are manufactured from a wide variety of carbonaceous materials, including sawdust and wood, peat, lignite, coal and petroleum coke, bone, coconut shells, and sugar. These substances are first converted to char by a carbonization process in which the raw material is slowly heated in the absence of air to temperatures up to 600°C in order to drive off volatiles and to increase the carbon content of the base material. The char is then activated by heating to a sufficiently high temperature (200 to 1000°C) in an oxidizing atmosphere (O_2, CO_2, H_2O) to cause extensive burnoff to noncarbon impurities. The resulting product is highly porous and has a high surface area (up to 2500 m^2/g). The nature of the surface and the porous structure of the carbon are determined in large part by the conditions of activation, i.e., time, temperature, and atmospheric composition, as well as by the characteristics of the base material.

The resulting activated carbon is largely crystalline and similar in structure to graphite. However, in contrast to the infinite layered planes of graphite, activated carbon is comprised of submicroscopic crystals or microcrystallites which are generally less that 100 A in diameter and 5 to 15 layers in thickness, arranged in a somewhat disorderly fashion. Most noncarbon impurities are located at the edges of these microcrystallites, while the microcrystallite faces or planes are highly structured and correspondingly free of impurities. During activation, macropores are formed within the carbon by burnout of the most reactive material on the microcyrstallite edges. Hence, the walls of the macropores, composed of microcrystallite edges, contain impurities and structural imperfections which impart a relatively high level of surface activity to the carbon. Micropores, on the other hand, are formed by burnout of carbon from the highly structured microcrystallite planes. The walls of the micropores, while constituting by far the major portion of the surface area of the carbon, do not contain the surface functional groups and degree of surface activity associated with the macropores. These surface functional groups are responsible for the acid/base characteristics associated with activated carbon.

Depending upon the conditions of activation, activated carbons may exhibit acidic or basic behavior. Carbons activated by exposure to oxygen at about 200 to 400°C (low-temperature or L-carbons) have acidic groups, while carbons activated at temperatures of 800 to 1000°C in a CO_2 atmosphere or in a vacuum and then exposed to oxygen at room temperature (high-temperature or H-carbons) exhibit basic properties. The acidic nature of the L-carbons is attributed primarily to the presence of carboxyl, phenolic hydroxyl, and quinone-type carbonyl groups. The nature of the basic groups, however, has not yet been satisfactorily established. In view of the acid/base characteristics of the surface functional groups, activated carbons may exhibit cation or anion exchange characteristics, depending upon the pH of the solution in contact with the carbon.

Boehm[3] devised a technique for determining the relative acid strengths of the surface oxides of L-carbons by titrating the carbons with bases of different strength, notably $NaHCO_3$, Na_2CO_3, NaOH, and $NaOC_2H_5$. While this technique is useful in attempting to identify the functional groups responsible for the acidic nature of the carbon surface, a more useful procedure for quantifying the acidity or basicity of the surface groups and thereby characterizing the ion exchange potential of the carbon is to titrate the carbon with various quantities of a strong base or strong acid. After a suitable contact time, the equilibrium pH is measured, and through the use of a charge balance expression, the surface charge of the carbon can be calculated as a function of pH. Figure 1 is a plot of the surface charge of three granular activated carbons as a function of pH, determined by titrating 10 g/ℓ carbon with NaOH and HCl and allowing a period of 1 week for equilibration. The HydroDarco 3000 is observed to exhibit the greatest negative surface charge at neutral pH values and would therefore be expected to exhibit the strongest cation exchange properties.

Huang and Ostovic[5] determined the zero points of charge (pH_{ZPC}) for three different carbons by titrating them with acid or base and measuring their electrophoretic mobility; they reported values for the pH_{ZPC} of Filtrasorb 400, Nuchar 722 and Nuchar C-190-N, of 7.1, 5.7, and 4.8, respectively. These values illustrate the different electrochemical properties of different types of activated carbons.

II. REMOVAL OF TRACE METALS BY ACTIVATED CARBON

Several potential mechanisms have been suggested to account for the observed ability of activated carbon to remove dissolved metals from aqueous solution. These include (a) cation exchange by the acidic surface oxides described above, (b) chemisorption, for example, with sulfide or oxide groups associated with the noncarbon impurities on the walls of the macropores, (c) precipitation of insoluble metal species within the carbon pores from an otherwise undersaturated bulk solution, the carbon surface serving as a nucleating agent, (d) adsorption of metals bound to organics in the water as organometallic complexes, and (e) chemical reduction of highly soluble, oxidized metal species to less soluble reduced forms (e.g., Cr(VI) to Cr(III)).

The significance of cation exchange as a removal mechanism is best illustrated in Figure 2, after Huang and Ostovic.[5] The uptake of Cd(II), in μmol/g carbon, increases with increasing pH; the extent of the uptake for different carbons increases as the cation exchange capacity of the carbon, which varies inversely with pH_{ZPC} (see above), increases. Hence, Nuchar C-190-N, with a pH_{ZPC} of 4.8, is more effective in adsorbing Cd(II) than is Filtrasorb 400 with a pH_{ZPC} of 7.1

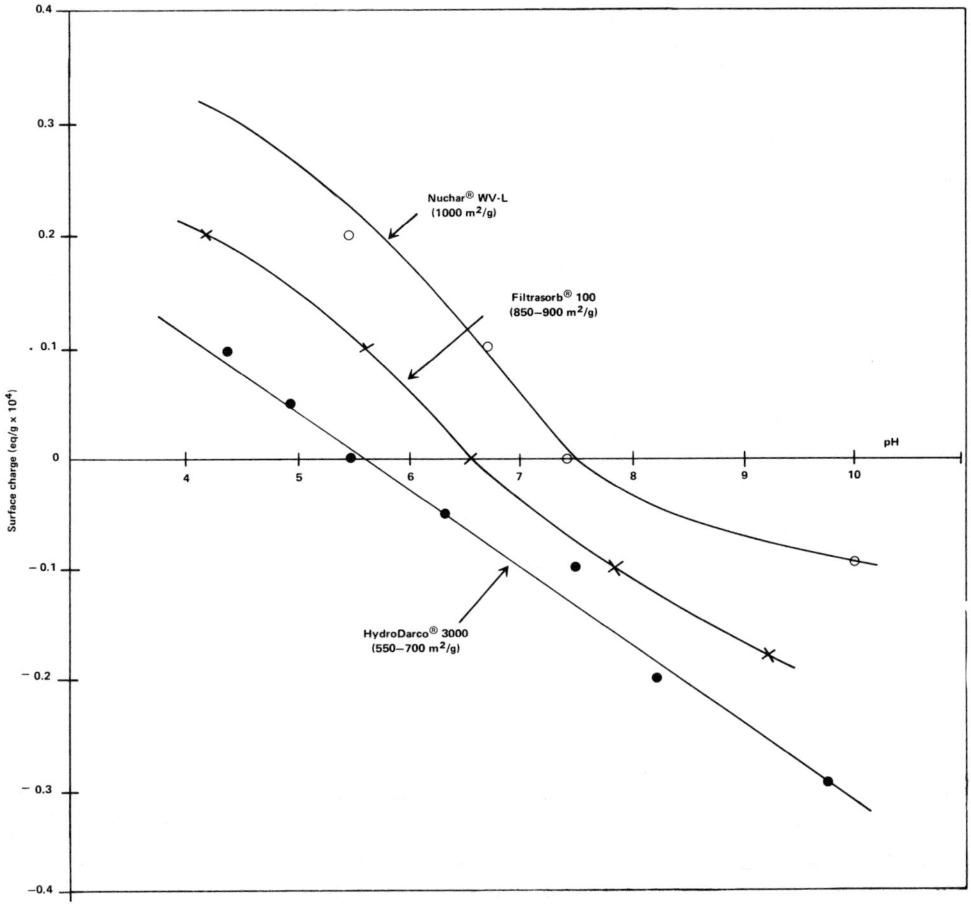

FIGURE 1. Effect of pH on surface charge of granular activated carbons. (After Peroutka, A. G. and Singer, P. C., unpublished research, University of North Carolina, Chapel Hill.)

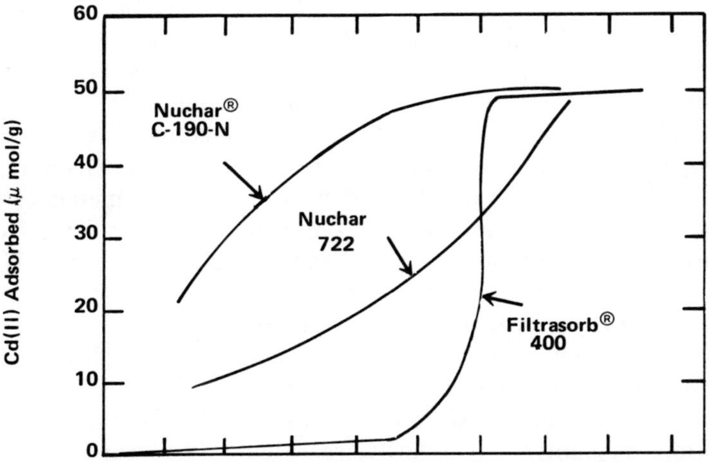

FIGURE 2. Effect of activated carbon type and pH on Cd(II) adsorption. (After Huang, C. P. and Ostovic, F. O., paper presented before the Division of Environmental Chemistry, American Chemical Society, New Orleans, La., March 22, 1977.

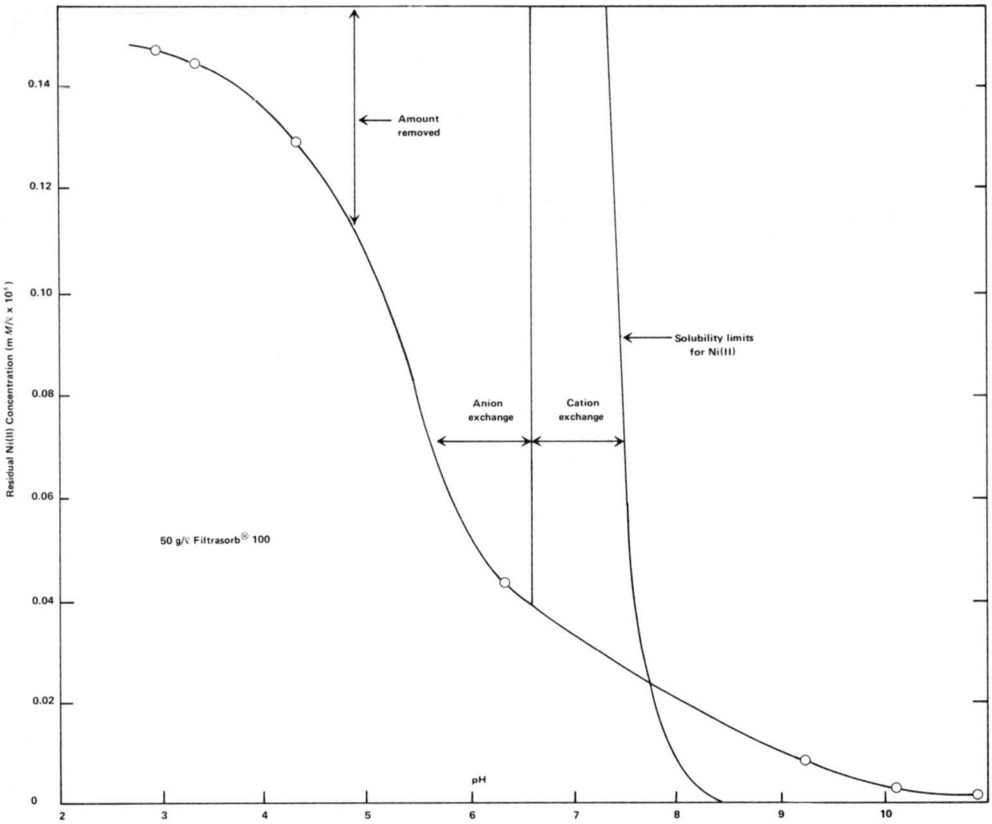

FIGURE 3. Effect of pH on Ni(II) adsorption by activated carbon. (After Peroutka, A. G. and Singer P. C., unpublished research, University of North Carolina, Chapel Hill.)

Figure 3, after Peroutka and Singer,[4] demonstrates that metal removal cannot be attributed to cation exchange alone. In model experimental systems, at pH values below 6.6, the zero point of charge of Filtrasorb 100, appreciable adsorption of Ni(II) is observed to occur, despite the fact that Ni(II) exists in these solutions predominantly as Ni^{2+} and that the net charge on the carbon is positive (see Figure 1). In addition, Nuchar C-190-N (pH_{ZPC} = 4.8) still adsorbs appreciable Cd(II) below pH 4.8 (see Figure 2). Specific chemical interaction between the metal and the carbon surface, or chemisorption, can be cited as an explanation for this observed behavior.

In column studies, Peroutka and Singer[4] have shown that metals can be taken up by different activated carbons in a manner opposite to that predicted from the cation exchange properties of the carbons. Figure 4 shows that the removal of Ni(II) increases in the order HydroDarco 3000 < Filtrasorb 100 < Nuchar WV-L, in spite of the fact that cation exchange capacity increases in the reverse order (see Figure 1). Effluent pH from the columns was observed to increase before breakthrough occurred; Peroutka and Singer attributed their observed metal removal to precipitation within the pores of the carbon. Comparisons[4] between the removal of Zn(II), Ni(II), and Cd(II) from separate solutions of equivalent molarity (see Figure 5) demonstrates that metal removal is inversely proportional to metal solubility, i.e., removal increases as the solubility of the metal decreases — thus, in order of increasing solubility, Zn(II < Ni(II) < Cd(II). It should be noted that these column experiments were carried out in small laboratory columns and no attempt was made to develop optimal loading conditions.

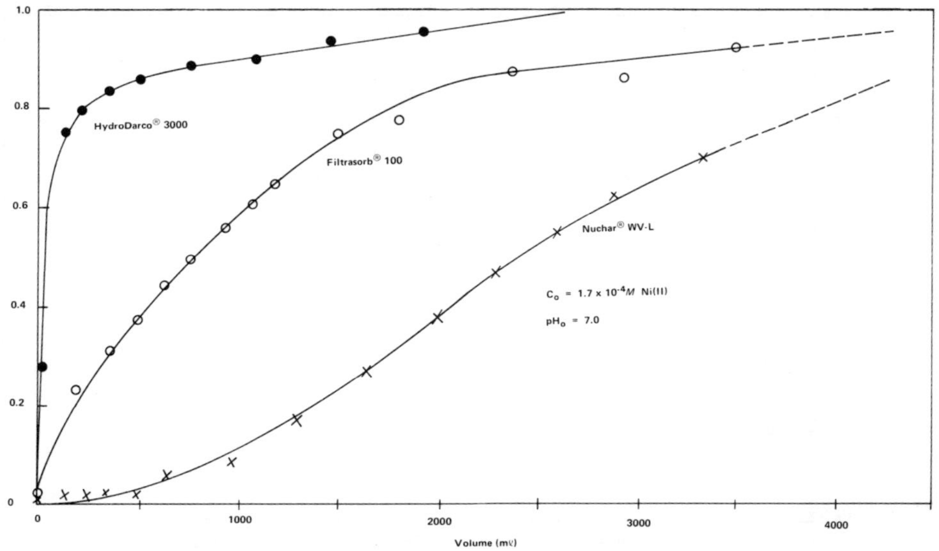

FIGURE 4. Removal of Ni(II) by granular activated carbon columns. (After Peoutka, A. G. and Singer, P. C., unpublished research, University of North Carolina, Chapel Hill.)

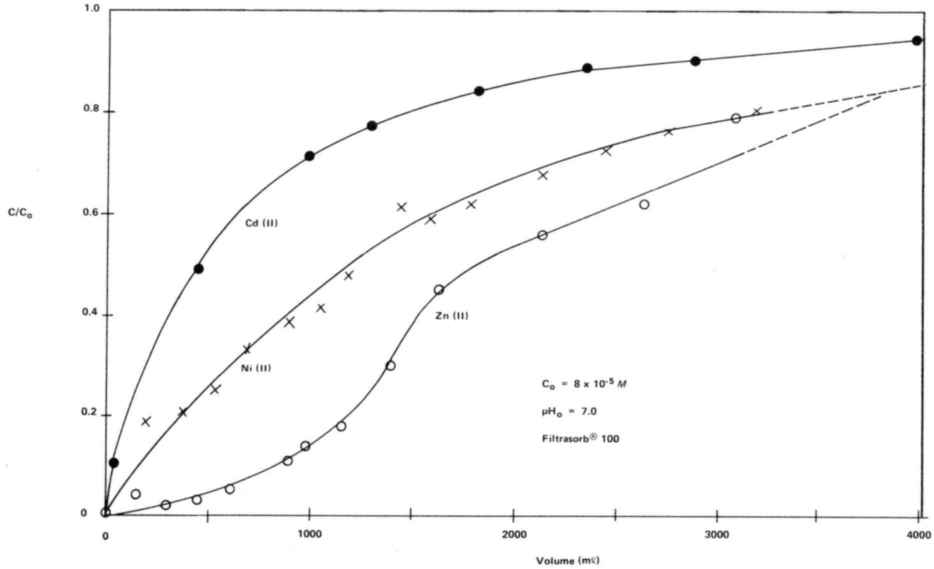

FIGURE 5. Removal of Zn(II), Ni(II), and Cd(II) by granular activated carbon columns. (After Peroutka, A. G. and Singer, P. C., unpublished research, University of North Carolina, Chapel Hill.)

The shape of the column performance curves reflect the high flow rate and unfavorable kinetics at which they were operated.

Organic species can significantly alter the extent of metal adsorption by activated carbon.[4,5,7] Figure 6 shows that EDTA, a strong metal-complexing agent, at a 1:1 *M* ratio relative to Cd(II), markedly improves Cd(II) adsorption, while phenol, which is essentially not a strong complex-former, exerts no apparent influence on the adsorption of Ni(II).

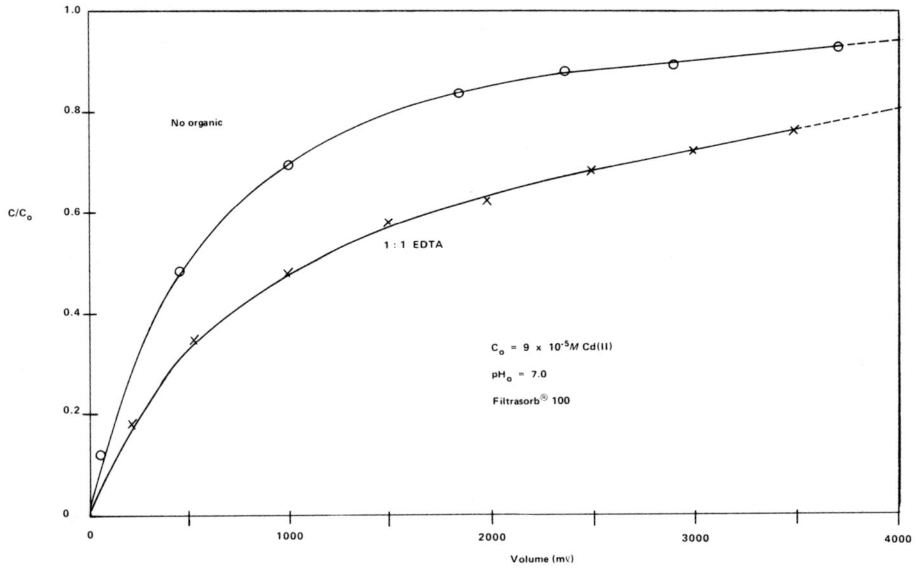

FIGURE 6a. Effect of EDTA (1:1 M ratio) on Cd(II) removal by granular activated carbon columns. (After Peroutka, A. G. and Singer, P. C., unpublished research, University of North Carolina, Chapel Hill.)

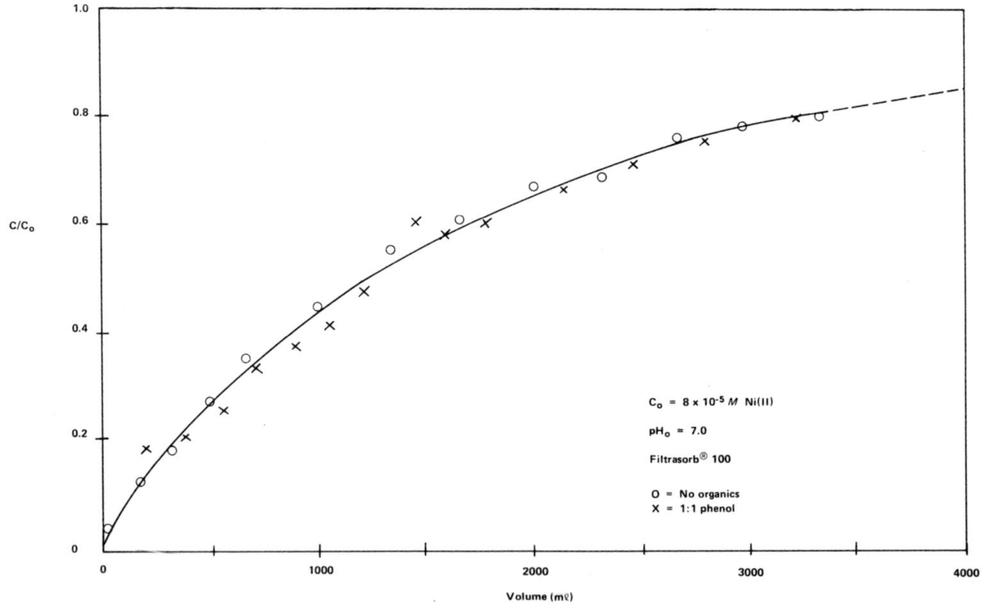

FIGURE 6b. Effect of phenol (1:1 M ratio) on Ni (II) removed by granular activated carbon columns. (After Peroutka, A. G. and Singer, P. C., unpublished results, University of North Carolina, Chapel Hill.)

These examples should emphasize that solution conditions and the chemical form of the metal can exert a marked influence on the uptake of metals by activated carbon.

III. MODIFICATIONS OF ACTIVATED CARBON FOR TRACE METAL REMOVAL

In view of the mechanisms by which activated carbon can adsorb trace metals from aqueous solution, it should be apparent that the capacity of the adsorbent can be enhanced by chemically treating the carbon to provide more surface functional groups and a higher degree of surface activity. For example, Humenick and Schnoor[8] "sulfurized" activated carbon by treatment with CS_2 and reported that Hg(II) was adsorbed stoichiometrically in response to the amount of sulfur on the carbon surface, suggesting chemisorption of Hg(II) by surface sulfide groups. Alternatively, treatment of the carbon with strong chemical oxidants, such as peroxydisulfate, permanganate, or hypochlorite, can be expected to provide additional surface oxides,[9,10] thereby increasing the cation exchange capacity of the carbon.

Nelson et al.[11] and Krause and co-workers[12] have developed a technique for improving metal adsorption on activated carbons by impregnating the carbons with metal oxides and sulfides. These inorganic adsorbents, by themselves, are difficult to prepare in a form suitable for adsorption column operation. However, with the carbon serving as a carrier, the metal-adsorbing characteristics of both the activated carbon and the metal oxides or sulfides can be utilized. Coconut charcoals containing as much as 6.9 mol Fe(III) as Fe_2O_3 per kilogram of carbon and 6.6 mol Zn(II) as ZnS per kilogram of carbon have been prepared, with a corresponding marked enhancement of Sr(II), Cu(II), Ag(I), Pb(II), Cd(II), and Co(II) adsorption reported.

Such techniques for modifying the chemical properties of these relatively inexpensive adsorbents and evaluating their potential for the removal of metals is worthy of further investigation. Lower-grade carbon adsorbents might also be similarly modified for this objective.

ACKNOWLEDGMENTS

The author would like to thank Alan G. Peroutka, a former graduate student at the University of North Carolina, Chapel Hill, for his experimental contributions which formed a basis for this paper and the School of Public Health of the University of North Carolina, Chapel Hill, for supporting his research. Appreciation is also due to Dr. C. P. Huang of the University of Delaware, Newark, for the use of his experimental results.

REFERENCES

1. **Mattson, J. S. and Mark, H. B., Jr.**, *Activated Carbon*, Marcel Dekker, New York, 1971.
2. **Snoeyink, V. L. and Weber, W. J., Jr.**, The surface chemistry of activated carbon: a discussion of structure and surface functional groups, *Environ. Sci. Technol.*, 1, 228, 1967.
3. **Boehm, H. P.**, Chemical identification of surface groups, *Adv. Catal.*, 16, 179, 1966.
4. **Peroutka, A. G. and Singer, P. C.**, unpublished research, University of North Carolina, Chapel Hill.
5. **Huang, C. P. and Ostovic, F. O.**, The Removal of Cadmium(II) From Dilute Aqueous Solution, paper presented before the Division of Environmental Chemistry, American Chemical Society, New Orleans, La., March 22, 1977.

6. **Singer, P. C.**, Chemical Processes for the Removal of Trace Metals From Drinking Water, *Proc. 16th Water Quality Conf.*, Vol. 71(108), University of Illinois, Urbana, 1974, 91.
7. **Thiem, L., Badorek, D., and O'Connor, J. P.**, Removal of mercury from drinking water using activated carbon, *J. Am. Water Works Assoc.*, 5, 447, 1976.
8. **Humenick, M. J., Jr. and Schnoor, J. L.**, Improving Mercury(II) Removal by Activated Carbon, *J. Environ. Eng. Div. Am. Soc. Civ. Eng.*, 100, EE6, 1249, 1974.
9. **Coughlin, R. W. and Ezra, F. S.**, Role of surface acidity in the adsorption of organic pollutants on the surface of carbon, *Environ. Sci. Technol.*, 2, 291, 1968.
10. **Coughlin, R. W., Ezra, F. S., and Tan, R. N.**, Influence of chemisorbed oxygen in adsorption onto carbon from aqueous solution, *J. Colloid Interface Sci.*, 28, 38b, 1968.
11. **Nelson, F., Phillips, H. O., and Kraus, K. A.**, Adsorption of Inorganic Materials on Activated Carbon, Proc. 29th Industrial Waste Conf., Engineering Bulletin, Engineering Extension Series, Purdue University, W. Lafayette, Ind., May 1974.
12. **Kraus, K. A., Nelson, F., Gilbert, J. S.**, Activated Carbons Filled with ZnS As Adsorbents, Proc. 31st Industrial Waste Conf., Engineering Bulletin, Engineering Extension Series, University, W. Lafayette, Ind., May 1976.

Chapter 22

CELLULOSE ION-EXCHANGE ADSORBENTS

M. C. Rowe

TABLE OF CONTENTS

I. INTRODUCTION

Early work on cellulose ion exchange was stimulated by the need to separate high molecular weight biochemical polyelectrolytes. Some work was carried out using cellulose itself, utilizing its small number of carboxyl groups, but more specialized ion exchangers were also developed and described in the early literature.[1,2]

The first major development in this field was the work of Sober and Peterson.[3,4] They developed exchangers based on cotton cellulose powder or sloka-floc wood cellulose with diethylaminoethyl (DEAE), carboxymethyl (CM), phosphoryl (P-C) and epichlorohydrintriethanol-amine cellulose (ECTEOLA-C) ionic functional groups (see Table 1). They were foremost in fractionating proteins, by utilizing the unique hydrophilic nature and porous structure of the cellulose matrix. In the separation of high molecular weight biological polyelectrolytes, ion exchange chromatography using these new exchangers was to prove more versatile than other available techniques such as electrolyphoresis or gel filtration. Guthrie and Bullock, in a review in 1960,[5] described exchangers derived from many different cellulose sources and illustrated the use of a cross-linking agent for the control of the swelling properties of the cellulose on introduction of the ionic groups.

A parallel development in the late 1950s was that of ion exchangers based on cross-linked dextran gel filtration materials. Diethylaminoethyl, carboxymethyl, and sulfomethyl derivatives were developed and extensively employed. More recently the development of ion exchangers based on regenerated cellulose (viscose)[6,7] have produced materials with potential for large scale utilization in biochemical processing and protein recovery from waste products. The new exchangers are cross-linked with epichlorolydrin and have DEAE, CM, sulfomethyl and quaternary ammonia ionic functional

TABLE 1

Ion Exchange Celluloses

$COOH-CH_2-O-$Cellulose	CM Cellulose	Weak acid cation
SO_3H-CH_2-O-Cellulose	SM Cellulose	Strong acid cation
PO_3H_2-O-Cellulose	P Cellulose	Weak acid cation
$(C_2H_5)_2>N-CH_2-CH_2-O-$Cellulose	DEAE Cellulose	Weak base anion
$CH_3-CH(OH)-CH_2-N^+(C_2H_5)_2-CH_2-CH_2-O-$Cellulose	QAE[a] Cellulose	Strong base anion

[a] Quaternary aminoethyl.

groups. Two types of material are available: viscose produced from fibrous cellulose before introduction of ion exchange groups, and viscose produced from fibrous cellulose ion exchangers. The two distinct products differ slightly in their affinity for some macromolecular species, but both are available in a granular physical form which has hydrodynamic properties superior to the fibrous cellulose analogs. In addition to the exchangers already described, the use of sulfonated wood pulp as an ion exchanger has been introduced on an industrial scale and shows promise in the treatment of sewage and industrial effluents. This material is a by-product of the wood pulping industry and hence has considerable manufacturing cost advantages.

Of the ion exchange products outlined, only the weakly acid exchanger CM-cellulose and the weakly basic exchanger DEAE-cellulose together with the wood pulp derivative, have been extensively used on a large scale. On a laboratory scale, all forms have proved useful, but in particular, DEAE and CM have been of general application.

II. CELLULOSE MATRIX PROPERTIES

A. Fibrous Cellulose

The matrix material consists of a random arrangement of microfibriles with a mean fiber length of 1000μm. This material resembles cotton-wool, but cellulose powders with a shorter fiber length (50 to 250 μm) are available. Flow rates tend to be poor by ion exchange standards (see Table 2), and attempts at flow rate improvements have resulted in microgranular forms which have a uniform particle size. The increase in flow rate is not significant, however, although the protein adsorption characteristics have been extended over a greater molecular weight range, as the material contains micropores which permit more easily the passage of macromolecules.

The polyglucose chains of fibrous cellulose are held together by hydrogen bonding; the interspaces are relatively large, although dense crystalline regions do exist. The matrix has favorable hydrophilic properties which help the approach of polar macromolecules, without significant electrostatic forces operating and subsequently denaturing the adsorbed species.

B. Regenerated Cellulose

In an attempt to produce a uniform and more porous matrix in a more acceptable physical form, exchangers were produced from viscose (regenerated cellulose). Two approaches to manufacture have been adopted. In one case, powdered regenerated cellulose produced from xanthation of wood cellulose is treated to provide ion exchange groups and cross-linked with epichlorhydrin.[6] In the second case, wood cellulose is converted to the ion exchange derivative and then regenerated with a lesser degree of cross-linking (epichlorhydrin).[7] Viscose is produced by dissolving fibrous cellulose in sodium hydroxide and reacting with carbon disulfide to form the xanthate. This product can then be regenerated by extruding into an acid bath. The material is then ground to produce the final product.

The physical properties of the regenerated cellulose vary due to the degree of cross-linking and method of manufacture. In general, the bulk density of the resin can be expressed as a swelling volume, 1 g of dry resin typically swelling to 4 or 5 mℓ. The wet density just exceeds one, when the resin is fully hydrated with little entrained air. The mean diameter of the pores is larger than protein molecular size, and protein capacities of greater than 1 g/g of exchanger have been recorded. The beneficial prop-

TABLE 2

Cellulose Exchanger Properties

Exchanger	Matrix type	Ionic functional group	Physical form	Flow rate
Fibrous cellulose e.g., Whatman®	Cotton Powder Solka-floc Polycell wood	Phosphoryl, CM, DEAE, ECTEOLA	Fibrous powders, microgranular	Slow ≃lmhr^{-1}
Regenerated cellulose Protion® or V.D.C.	Cross-linked viscose or cross-linked regenerated cellulose exchanger	CM, DEAE, SP, QAE,	Granular	6—12 mhr^{-1}
Sephadex®	Cross-linked Dextran®	CM, DEAE, SP, QAE	Powders or granules	Slow ≃1mhr^{-1}
Jorgensen® resins	Ligno-cellulose (wood pulp)	SO_3H aminated product	Wood particles	≃ 12 mhr^{-1}

After Goddard, J. E., in Proc. The Theory and Practice of Ion Exchange Conf, Society of the Chemical Industry, Cambridge, England, 1976.

erties associated with the fibrous cellulose matrix also hold for the regenerated product, and similar ion exchange groups can be attached, e.g., DEAE, CM, etc.

The accessibility and therefore efficiency of exchange in these products enables the number of ionic groups to be reduced, with 0.8 to 1.2 meq ionic residuals per gram of material being typical. The disparity with normal ion exchangers is even greater on a volume basis owing to the lower density of the exchangers. As the number of groups is lower, the possibility of irreversible bonding between the solute and resin phase is reduced. This property also protects the adsorbed protein from denaturation. The pK values for the ionic substituents is close to that in the free acids or amines, allowing the materials to be more generally applicable. In addition to pH considerations, the ion exchangers are very sensitive to ionic strength, which can be utilized in regeneration of the materials. A good regenerant for weakly acidic and weakly basic exchangers is 5% brine with a makeup of 2% alkali or acid, depending on the type of exchanger.

C. Ligno-Cellulose Wood Pulp Exchangers

These materials are based on sulfonated wood pulp, a by-product of the sulfite digestion of wood.[8] The materials are thus inexpensive and readily available in wood pulp producing countries. Sulfite liquors are used to solubilize lignin in the pulping process and release the cellulose fibers. If the process is interrupted, an ion exchanger, already sulfonated (cationic), is obtained. By using different pH values for the cooking liquor, it is possible to produce exchangers with different capacities. Of course, there is always an inherent weak acid capacity due to the carboxyl groups associated with the hemicelluloses present. Total exchange capacities of these materials lie in the range 0.1 to 0.4 meq/gram.

III. APPLICATIONS

A. Fibrous and Regenerated Cellulose Exchangers

A major application on a laboratory and pilot plant scale for the exchangers described in Section II. A and II. B (including Sephadex®) is ion exchange chromatography. The preparation, purification, and identification of high molecular weight biochemical polyelectrolytes is possible using these materials. Enzyme separation (shown in Figure 1), hormone separation, and serum protein separations and adsorptions have been achieved. The use of such materials is now standard in protein chemistry and the literature too large for a complete summary. Good reviews will be found in any ion exchange chromatography text book. Table 3 summarizes the salient points in this application. Widespread marketing of laboratory-grade regenerated cellulose derivatives should lead to increased utilization in the future (see Figure 1). The value of the regenerated cellulose matrix is well shown in the larger scale application of the materials. It is possible to work at much higher flow rates than the fibrous cellulose derivatives, and the kinetics are more specific.

Regenerated cellulose exchangers have been used on a large scale for the recovery of protein from waste streams: for example, serum protein from abbatoir waste, lactoalbumin from milk whey, and the purification of egg white protein.[9] The main conclusions to be drawn from this work are that even regenerated cellulose developed pressure loss problems under flow rates approaching 10 to 15 BV/hr (normal ion exchange flow rates), and that in the case of the abbatoir work, irreversibly bound "hemin" shortened the life of the exchanger. A chromatographic effect is observed using abbatoir waste, which has a profound effect on the run capacity of the exchanger. The effect is much less with poultry waste, as there is probably less hemoglobulin or other proteins with isoelectric points similar to the pH of the solution. Figures 2 and 3 show the effect of feeding poultry and slaughterhouse waste to a column containing DEAE regenerated cellulose exchanger.

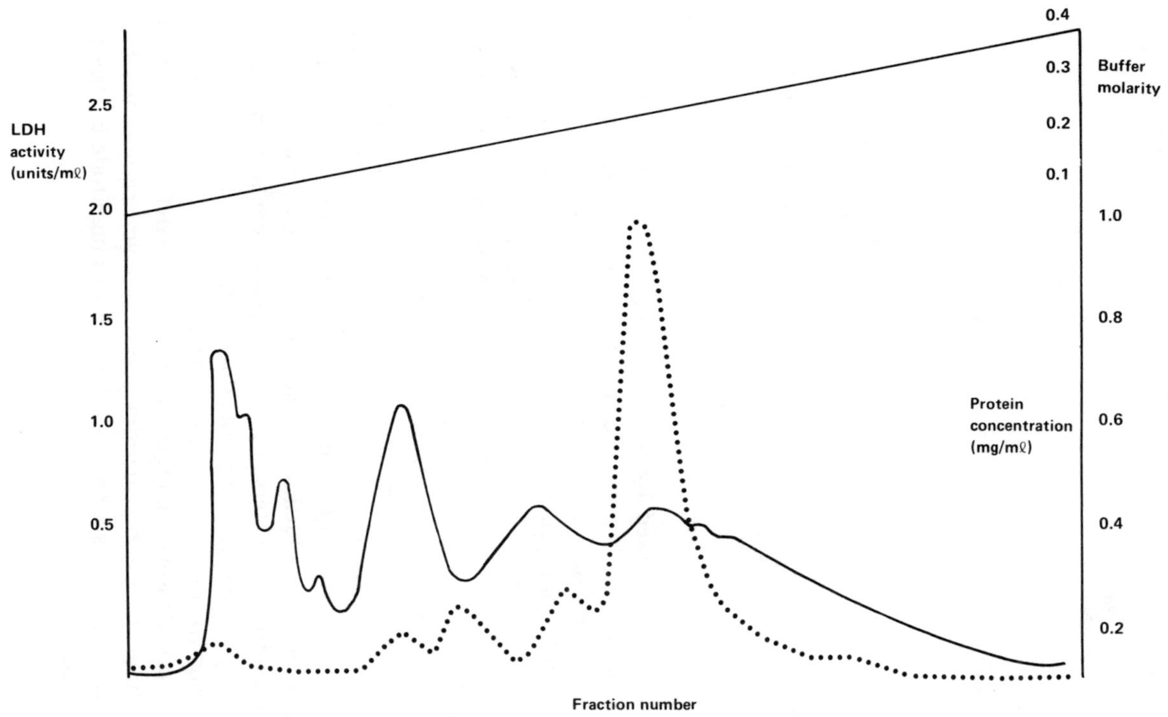

FIGURE 1. Separation of LDH isoenzymes (kidney extract). Resin, DEAE-viscose, 100 to 200 mesh; column dimensions, 30cm × 2cm; elution, concentration gradient; flow rate, 0.8mℓ/min.

TABLE 3

Cellulose Exchanger Capacities

Exchanger	Matrix type	Ionic forms	Total ion capacity (meq/g)	Protein capacity (mg/g)	Molecular weight range
Cell-O-P	Fibrous based only	NH_4^+	4	—	Peptides and nucleotides 8000 mol wt
CM	Fibrous and regenerated cellulose	Na^+/H^+ Na^+/H^+	0.6 1.0	1000 800	Proteins, enzymes, and cells 15,000—400,000 mol wt
DEAE	Fibrous based and regenerated cellulose	Free base Free base	1.0 1.0	500 800	—
ECTEOLA	Fibrous based only	Free base	0.5	—	Nucleic acids and residues

Note: Test systems used bovine serum albumin (anionic), 0.01 *M* Na Phosphate at pH 8.5; lysozyme (cationic), 0.01 *M* Na Acetate at pH 5.0.

One manufacturer has been prompted by this experience to utilize a stirred-tank reactor for such protein separations. The ion exchangers can be regenerated using alkaline or acid brine as previously described, and on adjustment of the pH, the protein is easily recovered from the column regenerant by heat coagulation. The returned protein was sufficiently undenatured to be used for animal feed supplement, as described by Grant.[10] The main economic consideration to emerge from the work was that due to the relatively high cost of the exchangers 4000 to 5000 U.S. $/t, the main deployment should be in recovery of high value proteins such as lactoalbumin (substitute egg white) and in waste treatment only if the economics or water recovery dictated it.

Outside of effluent and waste protein recovery, the regenerated cellulose exchangers have been used by the author for the large-scale production of bovine γ-globulin and selective separation of glucosidase. Outside the biochemical field, the removal of humic and fulvic acid from surface waters utilized for power generation has been achieved.[11] Work carried out by the author suggests that weak-base DEAE-regenerated cellulose exchanger has a significantly higher humic acid capacity than conventional macroporous polystyrene/acrylate ion exchangers, and within the definition of the two groups of regenerated exchangers, the product developed by regenerating the fibrous ion exchanger has humic acid capacities almost a full order of magnitude greater than its fellow viscose-based exchanger.

B. Ligno-Cellulose Resins

The major industrial application of the sulfonated wood pulp type exchangers is in waste water treatment. The work of Jorgensen has documented the use of the exchanger in a physicochemical treatment system. He reports the surprising result that the cellulo-lignin exchanger used upstream of a conventional macroporous exchanger facilitates greater organic removal capacity from the more conventional exchangers downstream[12-14] The ligno-cellulose/conventional ion exchange has been used on ab-

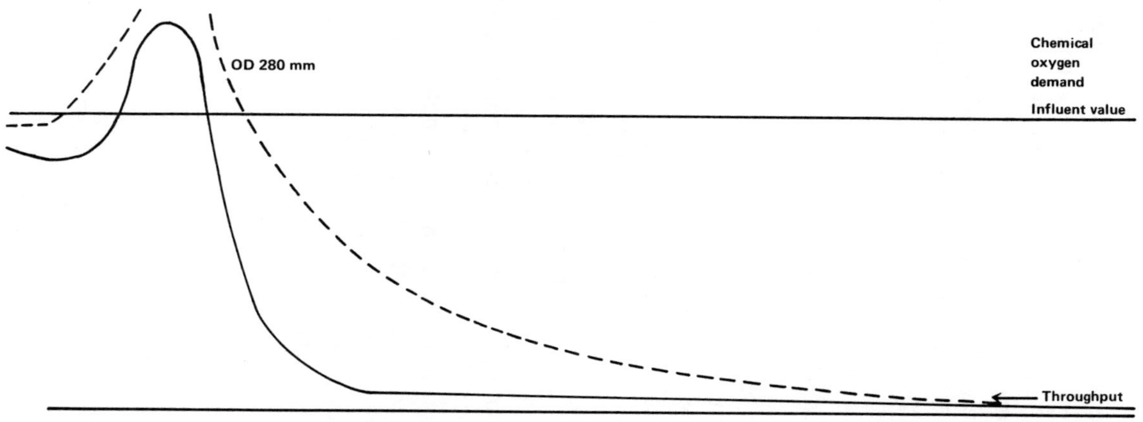

FIGURE 2. Meat works effluent (example of "chromatographic" effect). Resin, DEAE-viscose, 50 to 100 mesh; column dimensions, 0.1m × 1m; pH, 8.5; flow rate, 6 mhr^{-1}.

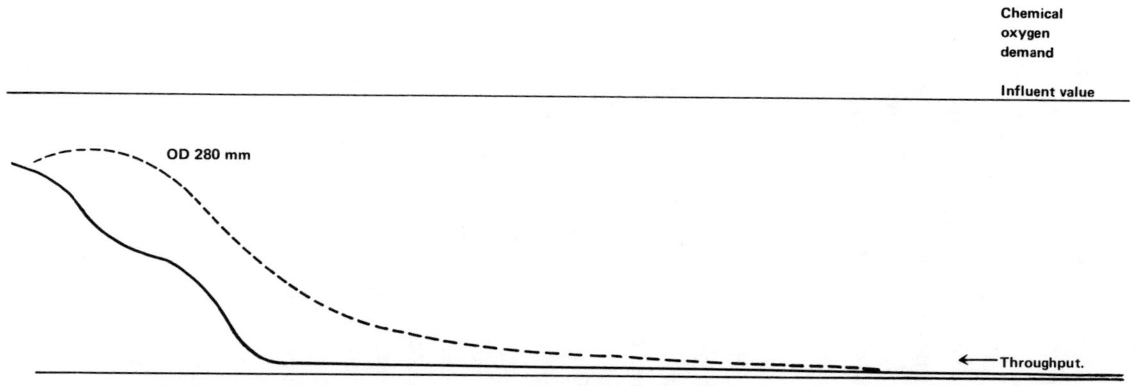

FIGURE 3. Poultry works effluent (example of "chromatographic effect). Resin, DEAE-viscose, 50 to 100 mesh; column dimensions, 0.1m × 1m; pH, 8.5; flow rate, 6 mhr^{-1}.

batoir, fish processing, and meat processing effluents, as well as on a larger scale in a sewage plant at Attran in Sweden.[15]

The ligno-cellulose material is, of course, a cation exchanger, and when compared with sulfopropyl-regenerated cellulose or carboxymethyl-regenerated cellulose exchangers, the protein capacity of the ligno-cellulose is between 100 and 150 mg/g (lysozyme capacity), and the regenerated celluloses are in the order of 500 to 1000 mg/g (lysozyme). These figures are in keeping with the quoted ion exchange capacities (0.27 to 1.0 meq/g). An interesting adjunct of the work on effluent treatment is illustrated in Table 4.

The feed stock to the ligno-cellulose material is a pretreated meat works effluent with an original chemical oxygen demand (COD) of ca. 3000 mg/ℓ. This was coagulated and allowed to settle, the treated liquor being removed by decantation and passed through the ligno-cellulose bed (0.25 meq/g exchange capacity).

This work opens the way for more definitive studies of bacterial purification using regenerated cellulose exchangers and bacteria and virus removal from treated effluents. These should be more effective than the wood pulp resins. Pioneer work has shown good results with tobacco mosaic and echo virus.

IV. CONCLUSIONS

The advent of regenerated cellulose resins has opened the way for industrial applications of cellulose exchangers. Work is proceeding on the removal of lactoalbumin from whey, globulin from bovine serum, and water recovery from waste streams on an industrial scale. More work should be carried out on humic acid adsorption and bacteria and virus control. Further industrial development will lead to some reductions in exchanger costs as well as innovative engineering to cope with the characteristics of a unique ion exchange material. At present, the cost differential offered to the industrialist by the wood pulp-based exchangers makes it possible for future development, particularly in the effluent treatment field. The main future application of regenerated cellulose exchangers on an industrial scale will probably come in the recovery of high-value protein constituents, but many potential applications presently available will be better served when development work has enabled the regenerated cellulose to be manufactured in a "harder" form which would be less compressible and therefore have more attractive hydrodynamic properties. The way now appears open for industrial manufacturers to produce an economically and technically viable product.

TABLE 4

Waste Water Effluent Properties

Aqueous sample	Before treatment	Before column	After column
Total coliform	>10^6/mℓ	>10^3/100 mℓ	Not detected
Aerobacteria genes	+[a]	Not detected	Not detected
Salmonella	+	Not detected	Not detected
Fermentation on meat peptones 3 days at 21°C	>10^6/mℓ	>10^2/mℓ	Not detected

[a] Indicates presence of sample but no count given.

REFERENCES

1. **Yackel, E. C. and Kenyon, W. O.,** Oxidation of cellulose by nitrite, *J. Am. Chem. Soc.,* 64, 121, 1942.
2. **McIntire, F. C. and Schenck, J. R.,** Polysaccharide acid esters as cation exchange media, *J. Am. Chem. Soc.,* 70, 1193, 1948.
3. **Sober, H. A. and Peterson, E. A.,**Chromatography of proteins on cellulose ion exchangers, *J. Am. Chem. Soc.,* 76, 1711, 1954.
4. **Peterson, H. A. and Sober, H. A.,** Chromatography of proteins. I. Cellulose ion exchange adsorbents, *J. Am. Chem. Soc.,* 78, 751, 1956.
5. **Guthrie, J. D. and Bullock, A. L.,** Ion exchange cellulose for chromatographic separations, *Ind. Eng. Chem.,* 52(11), 935, 1960.
6. **Grant, R. A. and Robins, J. C.,** Recovery of fats and proteins from wastewater, German Patent 2,210,379.
7. V.D.C., "Vistec" Cellulose Products Technical Publications, Viscose Group, Swansea, England.
8. **Wennerblum, B. A. and Jorgensen, S. E.,** British Patent 1,369,488, 1974.
9. **Jones, D.,** Protein recovery by ion exchange, *Process Biochem.,* 9(10), 17, 1974.
10. **Grant, R. A.,** Protein recovery from process effluents using ion exchange resins, *Process Biochem.,* 9(2), 11, 1974; *Effluent Water Treat. J.* 15(12), 616, 1975.
11. **Rowe, M. C.,***Effluent Water Treat. J.,* October 1975.
12. **Jorgensen, S. E.,** Danish Patent 6299/71.
13. **Jorgensen, S.E.,** *Water Pollut. Control,* 71, 210, 1972.
14. **Jorgensen, S. E.,** *Vatten,* 4, 332, 1968; also passim in *Vatten* 1971 to 1974.
15. **Jorgensen, S. E.,** Attran Sweden sewage plant, *Vatten,* 1, 36, 1973.

Chapter 23

MAGNETIC MICRO-ION-EXCHANGE RESINS

B. A. Bolto

TABLE OF CONTENTS

I. INTRODUCTION

The concept of employing ion-exchange resins containing ferromagnetic fillers was devised by Weiss and co-workers to overcome the difficult problem of handling very small resin beads.[1] It allows the rapid reaction rates of microresins to be exploited, an essential feature for the economic operation of certain inherently slow ion exchange reactions, such as those based on the thermal regeneration of mixtures of weak electrolyte resins.[2] More accessible ways of handling acidic and basic microresins of this type have been developed, utilizing the unique composite particles discussed in Volume I, Chapter 11, by Stephens. Nevertheless, historically, magnetic resins must be considered a spin-off from the thermal regeneration research program.

The magnetized forms of the microresins flocculate very strongly to give agglomerates with sedimentation rates comparable to those of normal-sized beads. However, the flocs readily disperse upon agitation; thus, the fast reaction rates associated with the small size of the primary particles can be achieved.[3] Magnetic resins therefore combine some of the handling characteristics of conventional resin beads with the rapid reaction rates of small particles. Their rather unique behavior makes possible the development of novel continuous contacting systems based on extremely simple equipment, well-suited to very large-scale operations. The evolution of magnetic polymers and recent developments in this nascent field are reviewed. The new technology has been patented, and exclusive licenses are held by ICI Australia Limited.

II. PREPARATION OF MAGNETIC RESINS

Magnetic resins have been prepared in a variety of physical forms. Homogeneous resins, containing magnetic material evenly distributed within the resin network, have been made by adding the magnetic material to conventional preparations, with the resins being obtained both as granules, via bulk polymerization (Figure 1a), and as spherical beads, by suspension polymerization (Figure 1b). When the magnetic resins are obtained as granules after grinding, some magnetic particles are left exposed on the surface of the resin; these are conveniently removed by appropriate washing with acid. Strong and weak electrolyte versions of homogeneous resins have been successfully prepared in cation and anion exchange forms.[3-6] An interesting variation on this theme is the postpreparative treatment of a sulfonic acid resin in the ferrous form under alkaline oxidation conditions to produce ferromagnetic domains.[7]

Heterogeneous resins of the shell type have also been developed, consisting of active polymeric chains grafted onto a core of magnetic polymer formed by embedding magnetic particles inside an inert cross-linked polymer.[8] Magnetic resins having an external shell of this type (see Figure 1c) are capable of rapid equilibrium, by virtue of their shell structure.[9,10] Typical dimensions of these resins are an overall diameter of 100 to 300 μm, with a shell thickness of 20 to 40 μm. Both weak and strong electrolyte versions of acidic and basic resins having shell structures have been reported.[4]

A number of magnetic materials have been investigated, with the emphasis on γ-Fe_2O_3 because of the small particle size of commercially available material, at about 0.1 μm, and its ready dispersion within the resin.[3] Ferromagnetic materials yielding stronger magnets, such as barium ferrite, are difficult to disperse because they are partly magnetic as supplied and are of much larger particle size, at about 2 to 3 μm. Other ferromagnetic materials which have been studied include Fe-Si and Fe-Cr,[6] while magnetite, synthetically precipitated Fe_3O_4, and a number of ferrites have also received attention. Some surface treatment of the magnetic material may be necessary prior to encapsulation by suspension polymerization methods.[3]

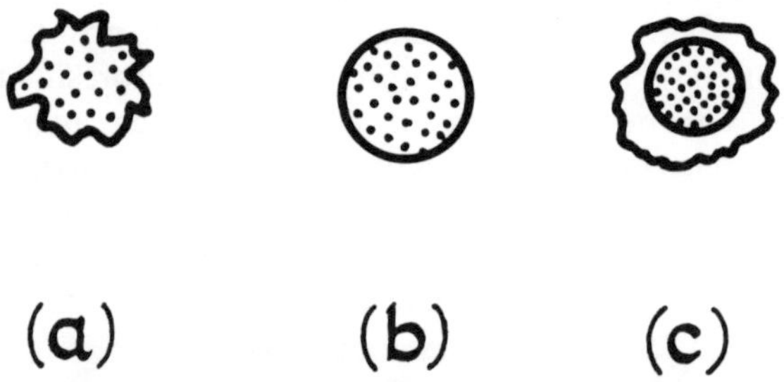

FIGURE 1. Types of magnetic ion exchangers. a. Homogeneous granules. b. Homogeneous beads. c. Heterogeneous shell resins.

III. PROPERTIES

The magnetic properties of the resins depend on whether the magnetic material is hard (barium ferrite), intermediate (γ-Fe_2O_3), or soft (Ferroxcube® a product of Philips Industries Limited). Magnetized hard ferrite resins behave as permanent magnets; once magnetized, they flocculate strongly and then are difficult to demagnetize. Soft ferrite resins containing Ferroxcube flocculate when in the presence of a magnetic field, but when withdrawn from the field, they retain little or no magnetism, and the floc therefore disperses. Intermediate resins containing γ-Fe_2O_3 retain substantial magnetism after being magnetized, and so flocculate outside the applied magnetic field. Although the resins can be partly demagnetized, complete demagnetization is difficult. Russian workers have classified their magnetic resins in terms of magnetic susceptibility. Only when the magnetic susceptibility was above $10{,}000 \times 10^{-6}$ cm^3/g were the resins, then in the strongly magnetic category, used in ion exchange procedures.[11]

There is a pronounced change in appearance when resins containing unmagnetized γ-Fe_2O_3 are magnetized by a brief exposure to a permanent magnet, achieved by passing the resin between the poles of a 1000 T horse-shoe magnet. Figures 2a and 2b show the development of chain structures after magnetization, and the formation of intense flocs. The magnetically flocculated samples show a dramatic increase in settling rate, as can be seen from the data in Table 1. These rates are about ten times faster on magnetization, and settling of the magnetized microresins is significantly faster than that of conventional resin beads of normal size.[3,12]

The settling rates are a function of the amount and type of magnetic material employed, of the size and structure of the composite particles (whether homogeneous or shell), and of the strength of the magnetizing field. The settling curves for a series of homogeneous weakly basic resins of varying magnetics content and bead size have been published.[3] Sedimentation of the magnetized resins is quicker the higher the magnetics content and the larger the bead size. A similar situation holds for the heterogeneous shell resins, which for the same magnetics content and overall particle size have a slightly slower settling rate than the homogeneous types because of the nonmagnetic layer of polymer surrounding the magnetic core. The decrease in the strength of the magnetized floc can be compensated for by having a higher magnetics content (see Table 1).

The high void volume of the magnetically flocculated resin, as illustrated in Figure 2B, is responsible for a quite unique feature of the system. Because of the presence of a higher proportion of water in the voids, compared to a conventional resin bed, the resin can be pumped directly without attrition, even from a settled bed. Pumping of the settled resin can hence be performed without the need for extra "hydraulic" water. This characteristic simplifies to an enormous degree the design of continuous contacting systems. It enables a flow of resin to be stopped and started without resin loss because of attrition and without difficulties arising from the settling of resin in transfer tubes or pipes.[4]

A final important property of the magnetized resin is that the flocs can be readily dispersed again in a zone of agitation, when the separate individual resin particles are obtained. These will react rapidly if the particles are of small size. The flocs will reform when the resin passes to a quiet zone. A comparison of the rate of reaction of a magnetic shell resin of overall particle size 100 to 300 μm, with that of conventional resin beads, 300 to 1200 μm in size, has shown that the smaller magnetic resin particles react more rapidly by at least an order of magnitude.[12] The resins were weakly acidic types, reacting with sodium bicarbonate in a simple stirred cell.

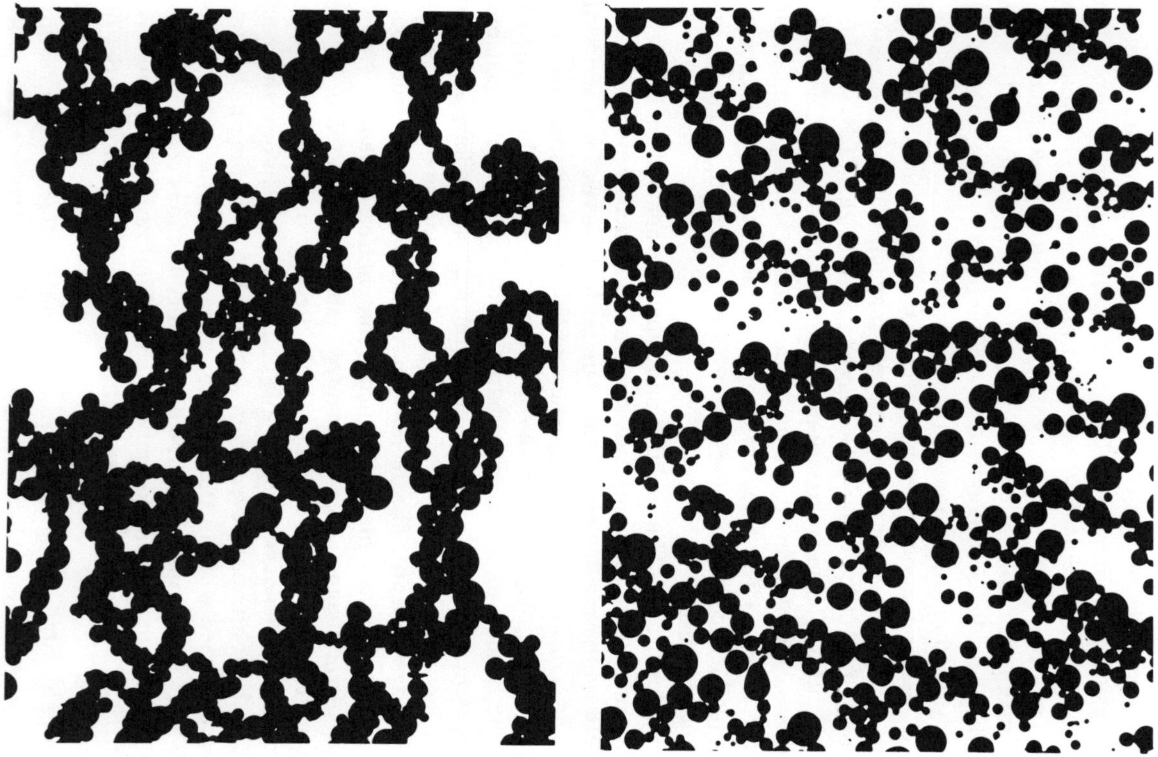

FIGURE 2. Homogeneous resin beads of about 50 μm diameter containing 20% w/w of γ-iron oxide. a. Unmagnetized. b. Magnetized. (From Blesing, N. V., et al., *Ion Exchange in the Process Industries,* Society Chemical Industry, London, 1970, 37. With permission.)

TABLE 1

Settling Rates of Magnetic Resins

Resin	γ-Fe_2O_3 Content (% w/w)	Particle size (μm)	Settling Time (sec)	
			Unmagnetized	Magnetized
Weakly basic, homogeneous	20	65—125	120	18
Weakly acidic, shell	50	100—300	100	10
Zerolit®[a]	Nil	300—1200	30	30

[a] A product of Diamond Shamrock (Polymers) Limited.

IV. CONTINUOUS CONTACTING PROCEDURES

The weakly acidic shell resin, based on polyacrylic chains grafted to a magnetic core, and containing about 50% γ-Fe_2O_3 by weight, has been used in studies of the dealkalization of hard, alkaline waters.[8] The reaction is as follows:

$$2RCO_2H + Ca^{++} + 2HCO_3^{\ -} \rightarrow (RCO_2^-)_2Ca^{++} + 2CO_2 + 2H_2O \qquad (1)$$

$$(RCO_2^-)_2Ca^{++} + 2H^+ \rightarrow 2RCO_2H + Ca^{++} \qquad (2)$$

The reaction is an ideal one for studies of the contacting method, as it can be performed as a simple, single-stage process.

A. Fluidized Beds

The first method tested for utilizing the resin in a truly continuous manner consisted of a fluidized bed adsorber, followed by a compact, moving-bed regenerator, with the two being linked by two resin pumps.[4,8] The resin is magnetized before use and is left in that state perpetually. Figure 3 shows a diagram of the assembly. The adsorber consists of a column with a conical base and a simple overlow launder with a magnetic grid along the lip. The feed water enters at the base of the cylindrical section and diffuses out through nozzles. Below the nozzles, an inverted conical baffle minimizes turbulence in the conical base of the column and allows resin to settle in this zone prior to being pumped out to the generator. The regenerator is equipped with inlet nozzles for wash water and acid and is fitted with a conical base to permit resin withdrawal. Loaded resin is pumped continuously from the base of the adsorber to the top of the regenerator at a fixed rate. The regenerated resin is pumped from the base of the regenerator and the resin flow is regulated by a photoelectric level-control at the top of the regenerator column. The system has been studied on both laboratory[8] and pilot-plant scales.[13,4] It can be seen that no screens, valves, or complex timing systems are involved and that there are no high pressures. A turbid water can be treated, and indeed, even the regenerant can be in the form of a slurry.

A significant loss of resin beads by elutriation from a highly turbulent fluidized bed can occur. It was found that such beads had agglomerated magnetically into small circlets of about six beads, which formed a magnetic circuit with such a small residual magnetic field that further growth into a large floc did not readily take place. In order to prevent loss of resin from this cause, it was necessary to install a simple magnetic trap in the overflow from the column. The circlets were either attracted to the magnets

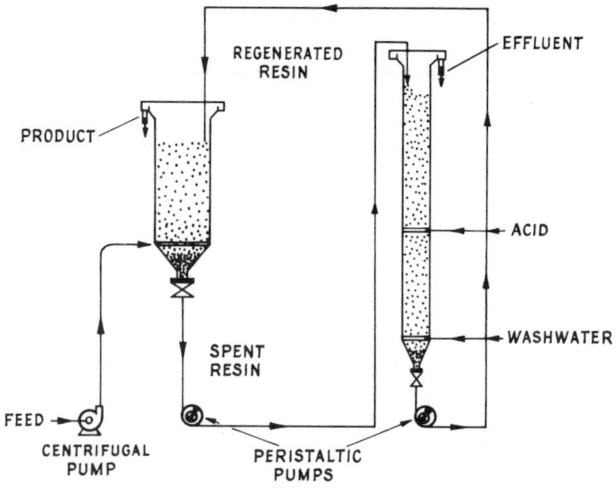

FIGURE 3. Flowsheet for a moving bed system, with a fluidized-bed adsorber unit. (From Bolto, B. A., Dixon, D. R., Eldridge, R. J., Kolarik, L. O., Priestly, A. J., Raper, W. G. C., Rowney, J. E., Swinton, E. A., and Weiss, D. E., *The Theory and Practice of Ion Exchange,* Society Chemical Industry, London, 1976, 1. With permission.)

themselves or reorientated so that they self-flocculated in a larger settling zone. In the pilot-plant trials, the loss of resin passing such a trap was monitored and found to be less than 1% per annum.[14,4]

The transfer of resin was accomplished in the laboratory apparatus by a pair of peristaltic pumps. Fresh resin suffered an initial loss of about 20% of its capacity in the first few hours of operation, but then there was no further loss of capacity over the 200-hr study.[8] The pilot plant was operated on a 3 m^3/day scale, using a variety of synthetic hard alkaline waters as feed, again with peristaltic pumps to transfer the resin.[13] It was then run using a very turbid feed of sewage origin, and after 300 hr of operation, there was a loss in capacity caused by the grinding action of quartz sand which had settled in the peristaltic pumps, so that it became necessary to replace these with eductors, whereupon no further loss in capacity from this cause was experienced over the 2500 hr the plant was operated. However, a slow chemical degradation has been observed arising from cleavage at the grafting point.[14]

The single-stage fluidized bed investigated for the adsorber system has proved to be eminently suitable for operation of a dealkalization process on a medium scale. For regeneration, an acid requirement of 106% of the theoretical amount was demonstrated. The synthetic feed waters utilized were made up to resemble hard, alkaline bore waters; data for these are available showing their dealkalization using conventional weakly acidic resins in fixed-bed and continuous modes employing normal commercial contacting equipment. On the 10,500 m^3/day scale utilized in the fixed-bed comparison, it was estimated from the magnetic resin pilot plant that the same result could be achieved with only 18% of the total resin volume.[13] Likewise, on a 170 m^3/day scale, comparison with a Boby CI (Asahi type) continuous plant showed that an appropriately scaled up magnetic resin system would require only 26% of the volume of resin needed by the commercial continuous contactor.[15] An example of the performances of the latter two systems is shown in Table 2 for the treatment of an underground bore water of the type obtained at the Australian Mineral Development Laboratories (AMDL) at Adelaide, in South Australia.[13,15] Comparable effects are obtained, the

TABLE 2

Dealkalization of Underground Water

	Fluidized bed of magnetic resin (mg/ℓ)	Boby CI plant (mg/ℓ)
Feed		
Ca^{++}	165	159
Na^{+}	197	175
HCO_3^-	366	357
Product		
Ca^{++}	64	46
Na^{+}	187	170
HCO_3^-	14	8
Effluent		
Ca^{++}	1960	1320
Na^{+}	203	245

greatest difference being the more concentrated effluent which is produced by the fluidized bed.

The major contrast between the systems is that resin transfer is truly continuous with the magnetic resin, rather than intermittent as in the Boby CI plant; also, the new method is not subject to pressure operation. In both comparisons, the magnetic system utilizes a significantly smaller and cheaper plant. Another vital advantage of the magnetic system is that an unclarified raw water may be treated without blocking up the bed. To prove this point, the fluidized-bed pilot plant has been operated under extremely arduous conditions, using as the feed primary settled municipal sewage, treated by lime flocculation and ammonia stripping, in a sidestream from a physicochemical treatment pilot plant.[16] The sidestream from the base of the ammonia stripping tower had an average turbidity of 37 nephelometric turbidity units (NTU), and a range of 5 to 135 NTU. The suspended solids included precipitated calcium carbonate, as well as organic material. The average chemical oxygen demand (COD) was 150 mg/ℓ. Some typical results of this study[14,17] are shown in Table 3. The system coped extremely well with the dirty unclarified feed, with most of the suspended solids going out in the product water. All of the suspended calcium carbonate was removed in the reactions occurring in the adsorption stage.

B. Mixer Settlers

The economic advantages of the fluidized bed for dealkalization using magnetic resins may be eroded in operation on a very large scale. On scaling up beyond medium size, the question of aspect ratio becomes significant, and the feed water distribution system becomes progressively more expensive. This is because of the axial mixing which occurs, requiring the bed to be two to three times deeper, relative to small-scale operations. Alternative systems such as mixer-settler arrangements involving continuous stirred-tank reactors, entrained pipe reactors, and even magnetic devices, all followed by resin-water separation systems, must be considered.[4]

Basic data have been obtained on resin settling rates and reaction rates for a wide variety of operating conditions.[4,18] The rate of reaction in the adsorption step has been found to be exponentially dependent on the initial loading of the resin, at constant concentration differences between the initial and equilibrium situations. This suggests that the reaction rate is not controlled by a film diffusion mechanism, but rather by diffusion through a loaded shell, into a shrinking core.[18] If this model is a reasonably accurate representation of the performance of these resins, then it can be inferred that the best way to take advantage of the high initial rate of reaction is to cycle the resin so rapidly that only the outer shell of the resin is reacted.

In order to provide a quick comparison of adsorber designs, continuous reactions were carried out using a stirred baffled tank. At a water-to-resin ratio of 25:1, starting with fully regenerated resin, complete dealkalization of the AMDL type bore water was achieved with a residence time of ca. 2 min. For an entrained pipe reactor, in which the resin is pumped cocurrently with the feed water to give a turbulent, plug flow situation, the required residence time was found to be only 45 sec, which is significantly smaller than for the stirred tank.[18]

A pilot plant has been operated on the 3 m^3/day scale, using the same lime-treated, ammonia-stripped sewage as feed, as described for the fluidized bed. The adsorber consisted of 60 m of 2.5-cm diameter flexible PVC tubing as the pipe reactor, which emptied into a settler, actually constructed from the former fluidized bed. The regenerator used in the early study was left unaltered in its incipiently fluidized bed form. Dealkalization studies in the new pilot plant showed that the same results could be achieved as with the fluidized bed adsorber, but using much less resin, circulated at a faster rate, and in simpler and cheaper equipment. A comparison of the adsorber systems contemplated so far is given in table 4, extrapolated to plants operating on a 4500 m^3/day scale.[17]

The more efficient utilization of resin in the pipe reactor may be ascribed to the better dispersion of the magnetic flocs in the highly turbulent flow within the pipe, compared to what is achievable in a fluidized bed, and even in a stirred tank. In the latter, the flocs are only broken up in the actual zone of shear and can re-form in any dead pockets within the tank. In the fluidized bed, axial mixing occurs; it is in the pipe reactor that minimum short-circuiting occurs because of plug flow.

Consideration has been given to simple settling tanks with rake mechanisms, inclined

TABLE 3

Typical Results for Dealkalization of Lime-treated Sewage Using Fluidized Bed of Magnetic Resin

		Concentration (mg/l)		
	pH	Ca^{++}	Alkalinity	SS
Feed	11.0	66	198	135
Product	5.2	22	12	120
Effluent	2.2	1580	—	—

TABLE 4

Comparison of Adsorber Units in Continuous Dealkalization on a 4500 m^3/day Scale

	Volume of adsorber unit (m^3)	Resin inventory (m^3)
Boby CI	3.8	2
Fluidized bed	16	2
Stirred tank/settler	6.5	0.85
Entrained pipe/settler	2.4	0.69

plate settlers, hydrocyclone systems and magnetic systems such as a drum separator or magnetized matrix filter. For flowrates of 4500 m^3/day, either the settling tank or the magnetic drum seems most feasible, but for larger flows, the settling tank becomes more economical. Such a tank would also serve as buffer storage for the treated water.[4] A flow sheet suitable for dealkalization but adaptable to all single-stage ion-exchange processes is shown in Figure 4. This incorporates a settler as part of the adsorption system and a second pipe reactor as the regenerator, followed by a vacuum drum filter to recover and wash the regenerated resin. The pipe reactor used as the adsorber for a 4500 m^3/day plant would comprise a pipe 22.5 cm in diameter and 100 m long. The extremely simple nature of the equipment is ideally suited to economical operation on a very large scale.

V. APPLICATIONS OF MAGNETIC RESINS

The dealkalization reaction has been studied in some detail, as a means of optimizing the contacting procedures. The handling methods devised are applicable to other ion exchange processes, with the proviso that multistaging may be necessary in certain cases.

A. Heavy Metals Removal

The weakly acidic shell resin used in the dealkalization work has also been employed for the adsorption of heavy metal ions. In particular, it has been used in the fluid bed mode for the adsorption of zinc ions from refinery effluents, in a three-column assembly.[19] The sodium loaded resin at pH 5 will remove zinc in preference to calcium ions from waters of composition similar to sea water, but containing in addition some 50 mg/ℓ of zinc. A product water containing less than 5 mg/ℓ of zinc was obtained, and after acid stripping, the effluent contained 370 mg/ℓ of zinc in a solution of pH 2. The third column is necessary to convert the acid form to the salt form by caustic treatment. Little more than the stoichiometric amount of reagents is required.

For certain applications in this area, it may be necessary to use more strongly acidic

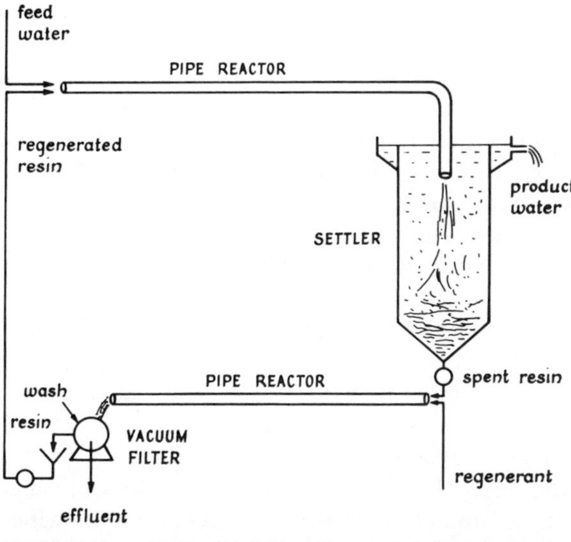

FIGURE 4. Generalized flowsheet based on entrained pipe reactors. (From Priestly, A. J., *J. Appl. Chem. Biotechnol.*, Submitted for publication. With permission.)

regeneration conditions. For the existing resin, solutions of pH 2 and greater can be tolerated without attack of the iron oxide within the cross-linked polymeric core. Precoating of the oxide is necessary to protect it from acid attack by solutions of pH less than 2. When a coating of a fatty acid modified polyamide is applied prior to encapsulation in the core, treatment of the final resin at pH 1 does not leach out iron.[1]

There would appear to be considerable potential for magnetic resins in the removal of heavy metals from waste waters, to allow both recovery of the metal, and reuse or harmless disposal of the waste water. The development of resins which are specific for certain metals is an obvious extension of this work. The economics depend very much on which metals are involved, but the approach has several advantages over the competing process of lime treatment in that operation is continuous, there is no sludge disposal problem, and there is the possibility of recovering the metal from the regeneration effluents.[19]

The use of magnetic anion exchange resins containing pyridine and pyridinium groups which are specific for the recovery of gold as the cyanide complex from industrial solutions and pulps has been reported.[20] The magnetic resins greatly facilitate the extraction of gold from such media.

B. Softening

Strongly acidic magnetic resins containing sulfonic acid groups are required for this operation. Such resins have been produced[4,6] but have not yet been utilized. Some softening experiments have been carried out with weakly acidic magnetic resins to estimate the efficiency of counter-current contact.[4]

C. Decolorization

Weakly and strongly basic magnetic resins perform well in the removal of color from paper industry effluents. The weakly basic variety is regenerated to the free-base form with caustic soda when the organics are stripped from the resin, as in the procedure using commercial resins.[21] The resin is then converted to the charged form with sulfuric acid, prior to another adsorption cycle. Although the kinetics are slower than for the adsorption of simple ions, it has been shown that there are substantial benefits to be gained by the use of magnetic resins for continuous decolorization, even with the requirement of multistaging. A very small resin inventory is required, and no preclarification is necessary so that large capital cost savings are possible.[14,29]

D. Clarification

Strongly basic magnetic resins of the shell type have been found to remove both suspended solids, such as clays, colloids, and algae, and dissolved organics, so that clarification and decolorization are achieved in the one step.[22] A small amount of alum is usually required as well, but not for algae. The positively charged polymer chains or "whiskers" on the exterior of the magnetic core are very effective at adsorbing the solid particles, which possess some negative charge, in a manner which is not possible with conventional strongly basic beads. The latter are quite rigid and allow only point-to-point contact, in contrast to the whisker resins, where the soft nature of the resin surface permits some immersion of the adsorbed particle and contact over a considerable proportion of its surface. Alkali is required for regeneration, but with algae, separation is readily achieved by agitation in a high frequency alternating magnetic field.

Unfortunately, the cost of the whisker resins renders the method uneconomical; however, a much simpler and cheaper inorganic version is showing exceptional promise in this area, where one of the major advantages is the production of a minimal amount of sludge.[23]

The combined clarification and softening of a turbid hard water has been demonstrated with magnetic whisker resins of the sulfonic acid type, which are preloaded with aluminium ions.[4] On mixing with waters containing sufficient calcium ions and alkalinity, the aluminium ions are released to the solution, where they react to destabilize and coagulate colloids in the usual way. It was hoped that in the regeneration step the aluminium hydroxide precipitated could be redissolved by addition of acid and the aluminium ions recycled by readsorption into the resin. However, this has not so far proved to be a feasible process, due to the unfavorable equilibrium on the resin of aluminium vs. hydrogen ions at the low pH necessary to redissolve the aluminium hydroxide. In trials to date, only about half of the aluminium can be readsorbed.

E. Sirotherm® Desalination

Magnetic versions of Sirotherm resins have been prepared by adding the magnetic material as additional particles to be contained within a composite bead,[2] as shown in Figure 2 of Volume I, Chapter 11, by Stephens. The overall composite particle size is reduced also, to give the sought-for rate advantage. Multistaging will be necessary, but even so, it is hoped that an inexpensive and reliable continuous process can be developed which would have significant gains over batch columns, in the form of a greater yield of product water, simpler control of alkalinity, easier exclusion of oxygen, longer resin life because of the shorter hot regeneration period, and most importantly, very significant energy economics.

VI. OTHER MAGNETIC POLYMERS

Magnetic polymers devoid of functional groups have been employed in two further water treatment processes, which make use of the high void volume of the magnetized materials. In hydrophilic form, homogeneous magnetic polymers perform well as magnetic filter aids, which can be recycled and reused, thus avoiding the disposal problem associated with spent one-shot filter aids.[24] As hydrophobic, vesiculated materials, they are under test for the removal of oil spills from harbor waters, where they are collected with up to 20 volumes of entrapped oil by simple magnetic means.[12]

Magnetic activated carbons have been prepared by heat treatment of carbons saturated with iron salts.[25] The product retains capacity for the adsorption of gold cyanide and allows ready separation of the loaded adsorbent from pulps by magnetic means. Magnetic activated carbons can be made which selectively adsorb only small organic molecules, macromolecules being excluded by the cross-linked polymeric network which binds the activated carbon and magnetic material into a composite bead.[26]

A further area related to magnetic resins is that of enzymes immobilized on solid magnetic supports. Enzymes have been coupled to iron oxide particles, with inert protein also present, by using glutaraldehyde as the coupling reagent.[27] Such magnetic enzyme resins, based on papain, have been maintained in a fluidized-bed reactor. A number of publications on this theme have appeared.[30,31,35] Magnetic bioaffinity materials have been used to purify enzymes.[32] Enzymes bound to magnetite[33] and antibodies bound to magnetic polymer spheres[34] will selectively attach some microbial cells.

VII. CONCLUSIONS

In addition to providing a practical means of handling rapidly reacting microparticles, which in itself means that the same job can be done using less resin, magnetic ion exchangers can be utilized in completely new contacting systems which are truly continuous in their operation, rather than intermittent, as in current commercially available systems. The equipment required is extremely simple and does not involve screens or high-pressure operation. As an added bonus, preclarification is not necessary. But the major benefit is that ion exchange reactions can be carried out to remove relatively small amounts of pollutants from very large volumes of water. If only a single equilibrium stage process is required, then the pipeline-separator system has a significant economic advantage over the fluidized bed. To achieve multistaging, multiple units would have to be installed. The great simplicity of the equipment means that operation on a 50,000 to 500,000 m^3/day scale can be readily envisaged.

ACKNOWLEDGMENTS

The author is indebted to the Society of the Chemical Industry, London, England, for permission to reproduce Figures 2 and 3, which are from References 3 and 4 and similarly to the *Journal of Applied Chemistry and Biotechnology* of Blackwell Scientific Publications Limited for the use of Figure 4, which is reproduced from Reference 18.

REFERENCES

1. **Weiss, D. E., Bolto, B. A., Willis, D., McNeill, R., and Ford, D. L.**, Ion Exchange Process, U.S. Patent, 3,506,378, February 2, 1971.
2. **Bolto, B. A. and Weiss, D. E.**, The thermal regeneration of ion exchange resins, *Ion Exchange and Solvent Extraction,* Vol. 7, Marinsky, J. A. and Marcus, Y., Eds., Marcel Dekker, New York, 1976.
3. **Blesing, N. V., et al.**, Some ion exchange processes for partial demineralization, *Ion Exchange in the Process Industries,* Society Chemical Industry, London, 1970, 371.
4. **Bolto, B. A., Dixon, D. R., Eldridge, R. J., Kolarik, L. O., Priestley, A. J., Raper, W. G. C., Rowney, J. E., Swinton, E. A., and Weiss, D. E.**, Continuous ion exchange using magnetic micro resins, Paper 27 in *The Theory and Practice of Ion Exchange,* Society Chemical Industry, London, 1976, 1.
5. **Lyubman, N. Y., Imangazieva, G. K., Svyadoshch, Y. N., Chatverikov, A. F., and Kim, Z. I.**, Synthesis of ferromagnetic ion exchangers by a granule polymerization method, *Tr. Nauchno Issled. Proektn. Inst. Obogashch. Rud. Tsvet. Metal.,* 5, 25, 1971: *Chem. Abst.,* 1977, 75873k, 1972.
6. **Imangazieva, G. K., Svyadoshch, Y. N., Lyubman, N. Y., and Lebedeva, A. D.**, Synthesis of ferromagnetic ion exchangers by a granule polymerization method, *Tr. Nauchnon. Issled. Proektn. Inst. Obogashch. Rud. Tsvet. Metal.,* 21, 1971: *Chem. Abst.,* 77, 75871h, 1972.
7. **Svyadoshch, Y. N., Imangazieva, G. K., Lyubman, N. Y., Kim, Z. I., and Shpil'Kova, A. E.**, Ferritization of strongly acidic sulfo cation exchangers, *Tr. Nauchno. Issled. Proektn. Inst. Obogashch. Rud. Tsvet. Metal.,* 5, 49, 1971: *Chem. Abst.,* 77, 7587m, 1972.
8. **Bolto, B. A., Dixon, D. R., Swinton, E. A., and Weiss, D. E.**, Continuous ion exchange using magnetic shell resins. I. Dealkalization — laboratory scale, *Ion Exch. Memb.,* 2, 1976, in press.
9. **Weiss, D. E.**, Counter current adsorption separation processes. II. Shell adsorbents, *Aust. J. Appl. Sci.,* 4, 510, 1953.
10. **Battaerd, H. A. J. and Siudak, R. V.**, Synthesis and ion exchange properties of surface grafts, *J. Macromol. Sci. Chem.,* 1259, 1970.
11. **Svyodoshch, Y. N., Masalov, A. M., Lyubman, A. Y., Imangazieva, G. K., and Kuznetsov, V. I.**, Behaviour of ferromagnetic ion exchange resins in a magnetic field, *Tr. Nauchno Issled. Proekt, Inst. Obogashch. Rud. Rsvet. Meta.,* 5, 157, 1971: *Chem. Abst.,* 77, 75940e, 1972.
12. **Bolto, B. A., Dixon, D. R., Eldridge, R. J., Swinton, E. A., Weiss, D. E., Willis, D., Battaerd, H. A. J., and Young, P. H.**, The use of magnetic polymers in water treatment, *J. Polymer Sci. Symp.,* No. 49, 211, 1975.
13. **Anderson, N. J., Dixon, D. R., and Swinton, E. A.**, Continuous ion exchange using magnetic shell resins. II. Dealkalization — pilot plant study, *Ion Exch. Membr.,* submitted for publication.
14. **Bridger, J. S. and Dixon, D. R.**, Continuous ion exchange using magnetic shell resins. IV. Pilot plant studies using unclarified feed.

15. **Melbourne, J. D. and Blesing, N. V.**, Processing of ground water for municipal use: dealkalization by continuous countercurrent ion exchange *Mech. Chem. Eng. Trans.*, Inst. Eng. Aust., MC4, 1958, 1968.
16. **Kolarik, L. O., Peeler, J. P. K., Pilkington, N. H., and Swinton, E. A.**, Physicochemical pilot plant, *Proc. 6th Natl. Convention,* Australian Water and Wastewater Association, Melbourne, May 1974, 345.
17. **Bolto, B. A., Dixon, D. R., Priestley, A. J., and Swinton, E. A.**, Ion exchange in a moving bed of magnetized resin, *Prog. Water Technol.*, 9, 833, 1977.
18. **Priestley, A. J.**, Continuous ion exchange using magnetic shell resins. V. Dealkalization — process development, *J. Appl. Chem. Biotechnol.*, submitted for publication.
19. **Dixon, D. R. and Hawthorne, D. B.**, Continuous ion exchange using magnetic shell resins. III. Treatment of effluents containing heavy metal ions, *J. Appl. Chem. Biotechnol.*, in press.
20. **Imangazieva, G. K., Lyubman, N. Y., Svyadoshch, Y. N., Shtrekker, G. P., Kim, Z. I., and Tabieva, L. K.**, Sorption of gold from cyanide solutions by ion exchange resins, *Tr. Nauchno. Issled. Proektn. Inst. Obogashch. Rud. Tsvet. Metal.*, 5, 113, 1971: *Chem. Abst.*, 77, 78105x, 1972.
21. **Anderson, L. G., Broddevall, B., Lindberg, S., and Phillips, J.**, A new color removal process: a field report, *Tappi*, 57, 102, 1974.
22. **Anderson, N. J., Bolto, B. A., Eldridge, R. J., Swinton, E. A., and Kolarik, L. O.**, Colour and turbidity removal with reusable magnetic particles. II. Flocculation with magnetic polymers, *Water Res.*, submitted for publication.
23. **Kolarik, L. O., Priestly, A. J., and Weiss, D. E.**, The Sirofloc Process for Turbidity and Colour Removal, Proc. 7th Natl. Convention, Australian Water and Wastewater Association, Canberra, October 1977, 143.
24. **Bolto, B. A., Cross, K. W. V., Eldridge, R. J., Swinton, E. A., and Weiss, D. E.**, Magnetic filter aids, *Chem. Eng. Prog.*, 71, (12), 47, 1975.
25. **Svyadoshch, Y. N., Kuznetsov, V. I., Sikolovskii, V. V., and Mostovaya, L. P.**, Ferritization of activated carbons, *Tr. Nauchno. Issled. Proektn. Inst. Obogashch. Rud. Tsvet. Metal.*, 5, 166, 1971: *Chem. Abst.*, 77, 78103V, 1972.
26. **Dixon, D. R., West, S., and Weiss, D. E.**, Selective magnetic adsorbents, *J. Appl. Chem. Biotechnol.*, submitted for publication.
27. **Gellf, G. and Boudrant, J.**, Enzymes immobilized on a magnetic support. Preliminary study of a fluidized bed enzyme reactor, *Biochim. Biophys. Acta,* 334, 467, 1974.
28. **Bolto, B. A., Dixon, D. R., and Eldridge, R. J.**, Graft polymerization on magnetic polymer substrates, *J. Appl. Polym. Sci.*, in press.
29. **Bolto, B. A., Priestly, A. J., and Sivdak, R. V.**, A Continuous Process for Decolorizing Caustic Bleach Plant Effluent Using a Magnetic Weakly Basic Ion Exchange Resin, Proc. 32nd APPITA Conf., Sydney, Australia, April 3 to 7, 1978.
30. **Robinson, P. J., Dunnill, P., and Lilly, M. D.**, The properties of magnetic supports in relation to immobilized enzyme reactors, *Biotechnol. Bioeng.*, 15, 603, 1973.
31. **Van Leemputten, E. and Horisberger, M.**, Immobilization of enzymes on magnetic particles, *Biotechnol. Bioeng.*, 16, 385, 1974.
32. **Dunnill, P. and Lilly, M. D.**, Purification of enzymes using magnetic bioaffinity materials, *Biotechnol. Bioeng.*, 16, 987, 1974.
33. **Horisberger, M.**, Immobilization of protein and polysaccharide on magnetic particles: selective binding of microorganisms by concanavalin A-magnetite, *Biotech. Bioeng.*, 18, 1647, 1976.
34. **Molday, R. S., Yen, S. P. S., and Rembaum, A.**, Application of magnetic microspheres in labelling and separation of cells, *Nature,* 268, 437, 1977.
35. **Munro, P. A., Dunnill, P., and Lilly, M. D.**, Non-porous magnetic materials as enzyme supports: studies with immobilized chymotrosin, *Biotech. Bioeng.*, 19, 101, 1977.

Chapter 24

APPLICATION OF MOLECULAR SIEVE ZEOLITES TO POLLUTION ABATEMENT

J. D. Sherman

TABLE OF CONTENTS

I. BACKGROUND

Molecular sieve zeolites are crystalline, hydrated aluminosilicates of (most commonly) Na, K, Mg, Ca, Sr, and Ba cations. The aluminosilicate portion of the structure is a three-dimensional open framework consisting of a network of AlO_4 and SiO_4 tetrahedra linked to each other by sharing all of the oxygens. Zeolites may be represented by the empirical formula

$$R^{n+}_{2/n}O \cdot Al_2O_3 \cdot X\,SiO_2 \cdot Y\,H_2O$$

In this oxide formula, X is generally equal to or greater than two since AlO_4 tetrahedra are joined only to SIO_4 tetrahedra; n is the cation valance. The framework contains channels and interconnected voids which are occupied by the cation and water molecules. The cations are quite mobile and can usually be exchanged, to varying degrees, by other cations.

It should be noted that the term "zeolite" properly refers to the *crystalline molecular sieve ion exchangers*. Unfortunately, some confusion in the literature has been created by the improper use of the term zeolite to include all inorganic materials exhibiting ion exchange capability. The accepted term for a synthetic *amorphous* aluminosilicate is "permutite." The molecular sieve zeolite cation exchangers discussed herein are of the *crystalline* framework aluminosilicate type and will be referred to by their scientifically correct designation "zeolite." In this chapter, the terms zeolite, molecular sieve, and molecular sieve zeolite are synonomous.

Although zeolites have attractive ion-exchange properties, they did not find significant use commercially as ion exchangers until the early 1960s. This was largely due to lack of availability and lack of knowledge of their properties. Both barriers disappeared during the 1950s and 1960s.

The discovery by R. M. Milton and co-workers[5] at Union Carbide that zeolites could be synthesized at convenient conditions (low temperatures and pressures) from reactive raw materials (e.g., freshly prepared aluminosilicate gels) led to the discovery of dozens of new zeolite structures. The fact that they could be synthesized by man assured their availability in commercial quantities in high purity, with reproducible properties. Extensive exploration initiated by Union Carbide in the late 1950s also resulted in the discovery in the western U.S. of many large natural deposits of certain sedimentary zeolites of sufficient quantity and purity for commercial use.

The first commercial ion-exchange uses were developed in the early 1960s by Ames et al.* for the processing of wastes from spent nuclear fuel. Later discovery (also by Ames et al.) that certain zeolites had excellent NH_4^+ ion selectivity led to the development in the late 1960s and early 1970s of the second significant commercial application: the removal of NH_4^+ from municipal waste water. At the present time, there are a number of other new applications of zeolite ion exchangers in various stages of development. Some of these are discussed briefly in the present chapter. The growing knowledge of their properties, together with growing needs for selective, stable ion exchangers in pollution abatement, water treatment, energy production, agriculture, aquaculture, animal nutrition, metals processing, and biomedical applications promise an exciting future in new applications of these unique materials.

The structure, chemistry, and use of zeolite molecular sieves has been broadly reviewed in a recent mongraph by Breck,[5] which includes a chapter extensively reviewing the theory, equilibria, and kinetics of ion exchange on zeolites. Applications of zeolites in ion exchange have also been described by Sherman.[15]

II. ION-EXCHANGE PROPERTIES OF MOLECULAR SIEVES

A. Typical Properties

Typical properties of the most common zeolite ion exchangers are summarized in Table 1. The ion-exchange capacity of zeolite ion exchangers is a function of their SiO_2/Al_2O_3 mole ratio, since each AlO_4 tetrahedron in the zeolite framework provides a single cation exchange site. Not all the exchange capacity is available to all ions. However, the majority of the total ion-exchange capacity is available to the most common cations.

Most common ions will exchange readily into most zeolites. However, ion sieving effects are observed with the zeolites having the smallest pore openings and with the

* These developments are reviewed in Reference 15.

TABLE 1

Typical Properties of the Most Common Molecular Sieve Zeolites

Zeolite type	Pore openings, hydrated form (Å)	Typical SiO_2/Al_2O_3 mole ratio	Typical maximum theoretical cation exchange capacity, Na^+ form, anhydrous (meq/g)
Analcime	2.6	4	4.9
Chabazite	3.7 × 4.2 2.6	4	4.9
Clinoptilolite	4.0 × 5.5 4.4 × 7.2 4.1 × 4.7	10	2.6
Erionite	3.6 × 5.2	6	3.8
Ferrierite	4.3 × 5.5 3.4 × 4.8	11	2.4
Mordenite[a]	6.7 × 7.0 2.9 × 5.7	10	2.6
Phillipsite	4.2 × 4.4 2.8 × 4.8 3.3	4.4	4.7
LINDE® A	4.2 into α cage 2.2 into β cage	2	7.0
LINDE® F	∼3.7	2	7.0
Zeolite HS (Hydroxysodalite)	2.2	2	7.0
LINDE® L	7.1	6	3.8
Large-port mordenite	6.7 × 7.0 2.9 × 5.7	10	2.6
LINDE® Omega	7.5	7	3.4
Zeolite P (LINDE® B)	3.1 × 4.4 2.8 × 4.9	3	5.8
LINDE® T	3.6 × 5.2	6.9	3.4
LINDE® W	4.2 × 4.4	3.6	5.3
LINDE® X	7.4 into α cages 2.2 into β cages	2.5	6.4
LINDE® Y	7.4 into α cages 2.2 into β cages	4.8	4.4

[a] The large pores are partially blocked in natural mordenite.

largest cations. For example, tetramethyl ammonium ions (6.9Å diameter) are excluded from clinoptilolite (4.4 × 7.2Å largest pore openings). It is clear that some ions are exchanged even though their hydrated ionic diameters are much larger than the pore openings; an exchange of solvent molecules must occur for such cations to diffuse through the pore openings. Ions which have anhydrous diameters ("crystal ionic diameters") which are too large are totally excluded. Ions may differ in their rates of exchange due to differences in the activation energy required to exchange solvent molecules.

In addition to some ion-sieving effects, zeolites commonly exhibit high selectivities for ion exchange among ions which will easily enter the zeolite pores. For example, LINDE® A zeolite provides a striking selectivity for Ca^{++} over Na^{++} compared with common organic resin, strong acid type cation exchangers.

B. Ion-Exchange Selectivities

Each zeolite provides a different pattern of ion exchange selectivity. The selectivity series for each of the most common zeolites have been reviewed by Sherman.[15] In general, common organic resin cation exchangers prefer ions of higher charge. This is also true of many of the zeolites (e.g., LINDE A and X zeolites). However, some zeolites show marked selectivity for some monovalent cations over common divalent cations. For example, LINDE® W exchanges NH_4^+ in marked preference to Ca^{++} and perfers Na^+ over Ca^{++}. Most of the zeolites exhibit selectivities for Ag^+, Tl^+, and Pb^{++} exchange and many also are selective for exchange of other nonferrous metal cations (e.g., Cd^{++}, Zn^{++}, Cu^{++}).

C. Variables Affecting Cation Exchange

The observed ion-exchange selectivities and loadings on zeolites are dependent upon the pH (H^+ is a competing cation), temperature, and aqueous solution chemistry. The competing cations, choice of solvent, presence of complexing agents, solution strength, and types of anions present can each alter the quality of the ion-exchange separation which can be achieved (via the effects of these variables upon the activities of the cations in solution as is also true in the case of organic resin ion exchangers).

Complexing of the cation can markedly alter its exchange properties. For example, Ag^+ exchange is very much favored over Na^+, but Na^+ is preferred over the $Ag(NH_3)_2^+$ complexed ion on LINDE X zeolite. Thus, regeneration of the exchanged zeolites may be accomplished using complexing agents in the regenerant solution. Also addition of complexing agents may allow zeolite ion-exchange separations (e.g., for selective non-ferrous metal exchanges) not otherwise obtainable.

III. ZEOLITE STABILITY

The molecular sieve zeolites have rigid, strong frameworks stable to high temperatures, oxidation/reduction, ionizing radiation, and not subject (as are many organic resin ion exchangers) to physical attrition due to osmotic shock. For the same reasons, the ion-exchange properties of the zeolites are relatively more constant and predictable over wide ranges of temperature, ionic strength, etc., than is often the case with other ion exchangers. Similarly, zeolite ion exchangers should not tend to adsorb organic molecules or ions and become "fouled" as readily as other ion exchangers.

Zeolites are also stable at elevated pH levels (e.g., pH 7 to 13) at which other inorganic ion exchangers (e.g., zirconium phosphates, etc.) tend to lose functional groups due to slow hydrolysis.[10] Zeolites are synthesized at elevated pH levels (e.g., pH 12 to 13+) and temperatures (e.g., 100 to 300°C) and are quite stable at conditions only slightly less severe than these employed during their synthesis.

The chief restriction in the use of zeolite ion exchangers is due to their limited acid resistance. Although some zeolites are stable at lower pH levels (pH$\sim$2), most zeolite ion exchangers should not be employed below about pH $\sim$ 4 to 5 except for very brief exposures. Operation at pH $>$ 6 is preferred.

IV. PRESENT APPLICATIONS

In the following sections, the present uses of zeolites in ion exchange will be discussed by type of application with emphasis on the unique properties of the zeolites employed in each. Recent advances in the development of improved zeolites for use in ammonium ion exchange are described. Later sections briefly describe other areas in which the unique properties of zeolite ion exchangers provide opportunities for new approaches to ion-exchange separations and purifications.

TABLE 2

Cesium and Strontium Radioisotope Recovery and Purification Using Molecular Sieve Zeolites

Service	Location	Zeolite
^{137}Cs/High-level wastes	Hanford, Wash.	LINDE AW-500
^{137}Cs/Rb, K, Na Purification of product from LINDE AW-500 column	Hanford, Wash.	Large-port mordenite
^{90}Sr, ^{137}Cs/Low-level waste water from fuel storage basin	National Reactor Testing Station, Idaho	Clinoptilolite
^{137}Cs/Evaporator overheads and miscellaneous waste water	Savannah River Plant, Barwell, S.C.	LINDE AW-500
^{137}Cs/Process condensate waste water	Hanford	Large-port mordenite

A. Ion Exchange of Cesium and Strontium Radioisotopes

Due to their stability in the presence of ionizing radiation and in aqueous solutions at high temperatures, inorganic ion exchangers offer significant advantages in applications involving the separation and purification of radioisotopes. Their low solubility over wide ranges of pH, particularly at high pH levels (up to about pH 13), together with their rigid frameworks and consequent dimensional stability and attrition resistance, have endowed the zeolites with properties which generally surpass those of the other inorganic ion exchangers. Exploratory studies of the ion-exchange properties of these materials revealed the high selectivities and capacities of several zeolites for cesium and strontium radioisotopes. Subsequently, several new processes were developed using these zeolites, as outlined in Table 2.[15]

B. Ammonium Removal from Municipal Waste Water Secondary Treatment Effluent

Ion exchange processes employing the zeolite clinoptilolite are in various stages of use, construction, or planning in several locations in the U.S. for removal of ammonium ions from municipal waste water[15] (see Table 3; Volume I, Chapter 19, by Mindler).

Although clinoptilolite performs quite well in this service, exchangers with higher capacity should provide significantly improved overall process performance in this and related applications. Studies begun in 1968 at Union Carbide Corporation led to the discovery that the synthetic LINDE® F zeolite is more effective than clinoptilolite in removing NH_4^+ from waste water.[4,5] Further laboratory column ion exchange tests have confirmed that LINDE® F zeolite provides higher cycled NH_4^+ loadings than does clinoptilolite.[17] However, it was found that the LINDE® F zeolite loses NH_4^+ exchange capacity on cycling due to a strong affinity for divalent ions (e.g., Ca^{++}), which are only partially removed during regeneration of the zeolite. Clinoptilolite also provides severely reduced breakthrough loadings when calcium salts are employed for regeneration. This can have an unfortunate effect on process economics to the extent that it requires use of NaOH instead of the less expensive $Ca(OH)_2$ for regeneration. These considerations suggested that still further improvements in the characteristics of the zeolite ion exchanger should be sought.

C. Improved NH_4^+ Exchangers

During the same time period, exploratory tests were made of other zeolites. It was

TABLE 3

Molecular Sieve Zeolite Municipal Waste Water NH_4^{++} Removal Installations

Location	Plant size (mgd)	Regeneration	Designers
Rosemount, Minn.	0.6	NaCl, NaOH With steam stripping to recover NH_4OH solution	Graver, Union, N.J.
UOSA,[a] Va.	22.5	NaCl, NaOH With air strip acid scrub to recover $(NH_4)_2SO_4$ (ARRP[b] process)	CH_2M-Hill, Bellevue, Wash.
Tahoe-Truckee, Cal.	6	ARRP Process	CH_2M-Hill, Bellevue, Wash.
Alexandria, Va.	54	(Planning stage)	Greeley & Hansen, Chicago, Ill.
Denver, Colo.	1	(Planning stage)	

[a] Upper Occaquan Sewage Authority, Va.
[b] Ammonia Regeneration and Recovery Process

discovered[16] that zeolites of the LINDE W-phillipsite-gismondine group provide superior NH_4^+ exchange characteristics in tests in simulated secondary effluents, even though these zeolites have lower maximum cation exchange capacities compared to the LINDE F zeolite.

Initial column ion exchange studies compared phillipsite with clinoptilolite and LINDE F Zeolites. These studies were later extended to include the zeolites phillipsite, LINDE B, and LINDE W, which represent three different members of the phillipsite-gismondine group of zeolites having related framework structures. In order to examine the NH_4^+ exchange capacities and selectivities of these zeolites, each was equilibrated with a mixed cation solution, and the resulting zeolite solid samples were anlayzed, providing the results shown in Figure 1. As may be seen, the phillipsite-gismondine type zeolites (phillipsite, LINDE B, and LINDE W) provide higher NH_4^+ capacities and selectivities compared to clinoptilolite. LINDE W zeolite provides the best performance of all, and it is superior to both clinoptilolite and LINDE F zeolites in both capacity and selectivity.

Cyclic column ion exchange tests were next made to compare the performance of LINDE W zeolite with that of clinoptilolite for NH_4^+ removal from municipal waste water. For regeneration, a mixed cation solution similar in composition to that use in the CH_2M-Hill process[21] was employed. The results, given in Table 4, reveal that the LINDE W zeolite provides about 2.5 times greater NH_4^+ exchange capacity compared to Hector, Calif. clinoptilolite.

Other studies[8] indicate that application of these zeolites for waste water NH_4^+ removal should provide substantial reductions in overall process costs. Current efforts are focused on the development and scale-up of manufacturing processes for these zeolites.[23]

V. FUTURE APPLICATIONS

A. Radioactive Waste Storage

As discussed earlier, zeolite ion exchangers are employed in separations of long-lived Cs and Sr radioisotopes. Radioisotopes can also be retained on zeolites for long-term storage by ion exchange onto the zeolite, drying the zeolite to prevent excessive pressure after the container is sealed, and sealing the containers by welding.[9]

Since zeolites contain alkali metal or alkaline earth oxides, alumina, and silica (ma-

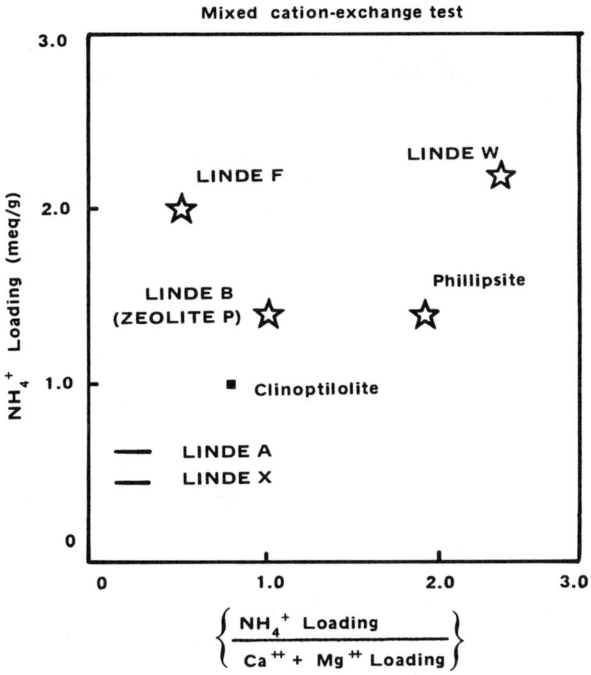

FIGURE 1. NH_4^+ loadings and selectivities of several molecular sieve zeolites.

jor constituents of many common glasses), heating to temperatures sufficient to cause destruction of the zeolite crystal structure can convert the zeolite to a glass. Leach rates for alkali and alkaline-earth elements from aluminosilicate glasses are extremely low (e.g., 10^{-7} g/cm²/day).[14] The chemical durability, low leach rates, and high thermal conductivity of glass combine to make this an ideal form for immobilizing radioactive wastes.

A new process developed at Sandia Laboratories employs a hydrous metal oxide type cation exchanger (e.g., Na Ti_2O_5H) to trap 90_{Sr} and other ratioisotopes from liquid wastes from fuel reprocessing, followed by a zeolite ion exchanger bed to trap the ^{134}Cs and ^{137}Cs.[1] Similarly, radioactive Cs, Sr, and Pu were sorbed on LINDE® AW-500 zeolite for final solidification in concrete or glass in studies at the Savannah River Plant, Barnwell, S.C.[7,24]

B. Molecular Sieve Zeolite Enhanced Biological Nitrification

The removal of NH_4^+ from municipal waste water by use of zeolite ion exchangers in a physical-chemical treatment process involving alternating cycles of loading and regeneration was discussed earlier. In such cycles, the zeolite is regenerated by mass-action or chemical driving forces.

The NH_4^+ loaded on the zeolite may also be removed by biological nitrification

$$NH_4^+ + 2O_2 \xrightarrow[\text{Bacteria}]{} NO_3^- + 2H^+ + H_2O$$

as is practiced in advanced waste water treatment. In effect, the addition of a zeolite to the activated sludge will impart selective NH_4^+ exchange capabilities to the sludge, thus improving its ability to remove NH_4^+ from the waste water. The conversion of NH_4^+ by nitrifying bacteria will regenerate the zeolite. Such "zeolite-enhanced biological nitrification" has been demonstrated.[18,19] The ability of the zeolite to pick up NH_4^+

TABLE 4

Column Ion Exchange NH_4^+ Removal Using Molecular Sieve Zeolites

	Zeolite	
	Hector, California clinoptilolite	LINDE W development sample
After pH 8.7 regeneration, water wash	180 BV	453 BV
After pH 7 regeneration, water wash	177 BV	481 BV

Note: Feed = 20 BV/hr of 15 NH_4^+ (as N), 40 Na^+, 10 K^+, 50 Ca^{++}, and 1 mg/ℓ Mg^{++}; regenerant = 7900 Na, 400 K, and 4400 mg/ℓ Ca.

during peak load periods and subsequently gradually release it may provide higher overall nitrification rates and improved ability to handle shock loads.

C. NH_4^+ Removal from Industrial and Agricultural Waste Waters

In addition to treatment of municipal waste water, it is anticipated that zeolites will find use in the removal of NH_4^+ from industrial and agricultural waste water streams. Here the availability of several different zeolites, each having NH_4^+ ion-exchange selectivity, with different properties offers the possibility of selecting the optimum exchanger for a particular service on the basis of its ability to selectively remove NH_4^+ in the presence of different competing cations.

D. Aquaculture

Ammonia is extremely toxic to aquatic animals. In closed systems (e.g., aquariums) and when extensive water reuse is practiced in high-density fish culture (hatcheries, fish farming), the ammonia released directly by the fish, from their other nitrogenous wastes such as urea, and from bacterial deamination of protein in food and wastes, will quickly reach toxic concentrations if not removed. Microbiological filters may be used for this purpose. However, nitrifying bacteria are easily inhibited or killed by various stresses (low temperature, sulfides, methanol, heavy metals, antibiotics used to control disease outbreaks, shock loads, etc.). Toxic levels of ammonia may be quickly reached before the biological filter operation can be reestablished to required levels. For these reasons, a number of investigators[11,12] have studied NH_4^+ removal by ion exchange on zeolites as an independent standby backup system (for emergency use when upsets occur, or during treatment to remove ammonia which escapes the biological filter), or for reliable ammonia removal in lieu of biological filters. Braico[2] concluded that zeolite ion exchange offers the advantages of lower cost, higher removals, a chemical process which is more controllable than existing biological processes, and lower land area requirements, for ammonia removal for reuse of fish hatchery waters.

Johnson and Sieburth[6] studied the use of LINDE AW-500 and clinoptilolite zeolite ion exchangers for ammonia removal from distilled water, from an active, closed system for Chinook salmon, and from artificial sea water of 5 to 25% salinity. They concluded that a zeolite ion-exchange column is desirable as a secondary or backup system to biological filters for use in low-density, closed aquaculture systems in which the high salinity reduced the effectiveness of the zeolites tested. They noted that in fresh water systems it is feasible to use ion exchange alone. By total removal of the ammonia, zeolite ion exchange also avoids the further complication of nitrate buildup resulting from the biological oxidation of ammonia.

Slone et al.[20] demonstrated the utility of zeolite (clinoptilolite) ion exchange in removal of ammonia escaping from a biofilter in a closed "silo" type fish culture system.

Torii[22] reviewed uses of natural zeolites (clinoptilolite and mordenite) in Japan and estimated that 5 to 10 million pounds per year of clinoptilolite are employed in removal of ammonia from aquaculture ponds or tanks. Spreading 500 kg clinoptilolite per 1000 m^2 of pond area successfully prevented problems from toxic ammonia generation (from sludges in the bottom of fish ponds), allowing culturing of colored carp to begin 4 days later. Similarly, clinoptilolite has been employed in ammonia removal from circulating water in fish breeding tanks.

E. Agricultural and Horticultural Applications

The agricultural uses of zeolites for feeding of swine and poultry, use in odor control in poultry houses, use of zeolites as soil conditioners, in fertilizers, as carriers of fungicides and pesticides, and as NH_4^+ ion exchangers to prolong the life of cut flowers, have recently been reviewed.[13,15,22]

F. Heavy Metals Removal, Recovery, and Separation

As reviewed elsewhere,[15] many zeolites exhibit high selectivities for various heavy metals. Therefore, zeolite ion exchangers are under consideration for use in recovery of precious and semiprecious metals and for removal of heavy metals from industrial and metals processing waste waters.

Because of their availability (especially in Japan), the zeolites clinoptilolite and mordernite have been studied for heavy metals removal from waste waters (especially Cd, Cu, Pb, and Zn). The very high selectivities of several zeolite ion exchangers for Ag^+ also suggest their use for the recovery of Ag^+ from waste waters.

Separations and purifications of nonferrous metals may also be accomplished by zeolite ion exchange, for example the unique separation of Co^{++} and Ni^{++} on LINDE A zeolite.[3] Many other separations of nonferrous metals are also possible. Separations of both free and complexed ions may be accomplished,[3,5] suggesting that zeolite ion exchangers may provide unique new separations and purifications in the processing of nonferrous metals.

VI. SUMMARY

Molecular sieve zeolite ion exchangers provide many unique combinations of cation exchange selectivity, capacity, and stability not available in other inorganic or organic ion exchangers. New applications of ion exchange employing the special properties of zeolites have been developed in separation of radioisotopes and in waste water ammonia removal. New zeolite ion exchanger products with still higher NH_4^+ exchange selectivities and capacities have been discovered and developed for use in waste water treatment, agriculture, aquaculture, biomedical, and other applications.

New applications of zeolite ion exchangers in nuclear fuel reprocessing and disposal, heavy metals removal, and other industrial and agricultural uses are under development and other potential uses in metals recovery and separations and purifications of nonferrous metals are being explored. Applications have grown rapidly with increasing awareness of the properties and the imaginative consideration of the potential uses of molecular sieve zeolites in ion exchange for pollution abatement and other uses.

REFERENCES

1. **Anon., New process consolidates radioactive wastes,** *Chem. Eng. News,* 54, 32, 1976; New processes and technology alert: consolidating radioactive wastes, *Chem. Eng.,* 83(15), 158, 1976
2. **Braico, R. D., M. S.,** thesis, Montana State University, Butte, 1972.
3. **Breck, D. W.,** *J. Chem. Educ.,* 41, 678, 1964.
4. **Breck, D. W.,** U.S. Patent 3,723,308, March 27, 1973.
5. **Breck, D. W.,** *Zeolite Molecular Sieves, Strucuture, Chemistry, and Use,* Wiley-Interscience, New York, 1974.
6. **Johnson, P. W. and Sieburth, J. M.** *Aquaculture,* 4, 61, 1974.
7. **Kelley, J. A., Hale, W. H., Stone, J. A., and Wiley, J. R.,** *A.I.Ch.E. Symp. Ser.* 72, (154), 128, 1976.
8. **Kepple, L. G.,** *Water and Sewage Works,* 121(4), 42, 1974.
9. **Mercer, B. W. and Schmidt, W. C.,** Atomic Energy Commission Accession No. 14466, Report No. RL-SA-58, Washington, D.C., 1965.
10. **Larsen, E. M. and Vissers, D. R.,** *J. Phys. Chem.,* 64, 1732, 1960.
11. **Liao, P. B. and Mayo, R. D.,** *Aquaculture,* 1, 317, 1972.
12. **Liao, P. B. and Mayo, R. D.,** *Aquaculture,* 3, 61, 1974.
13. **Mumpton, F. A. and Fishman, P.,** A Brief Review of Agricultural Uses of Natural Zeolites, paper presented at Zeolite '76 Conf., Tucson, Ariz., June 1976.
14. Oak Ridge National Laboratory, Chemical Technology Division Annual Progress Report, period ending March 31, 1975, ORNL-5050, October 1975.
15. **Sherman, J. D.,** Ion Exchange Separations with Molecular Sieve Zeolites, paper presented at the 83rd Natl. Meet., American Institute of Chemical Engineers, Houston, Texas, March 20 to 24, 1977.
16. **Sherman, J. D. and Ross, R. J.,** Separation of Ammonium Ions from Aqueous Solutions, German Patent 2,531,338, February 12, 1976.
17. **Sherman, J. D. and Ross, R. J.,** Union Carbide Corp., unpublished results, 1971, 1973.
18. **Sims, R. C.,** *Environ. Sci. Eng. Notes,* 9, 2, 1972.
19. **Sims, R. C. and Little, L. W.,** *Environ. Lett.,* 4(1), 27, 1973.
20. **Slone, W. J., Turner, P. R., and Jester, D. B.,** Proc. 54th Annual Conf., Western Assoc. of State Game and Fish Commissioners, Alburquerque, N.M., July 16—19, 1974, pp. 351—394.
21. **Suhr, L. G. and Kepple, L.,** Deisgn of a Selective Ion Exchange System for Ammonia Removal, paper presented at the American Society of Civil Engineers Environmental Engineering Division Conf., Pennsylvania State University, July 1974
22. **Torii, K.,** Utilization of Natural Zeolities in Japan, paper presented at Zeolite '76 Conf., Tucson, Ariz., June 6 to 14, 1976.
23. U.S. Environmental Protection Agency, Design Manual for Nitrogen Control, 1975, chapter 7.
24. **Wiley, J. R. and Wallace, R. M.,** Savannah River Laboratory Report DP-1388, Atomic Energy Commission, Washington, D.C.

APPENDIX A

BIBLIOGRAPHY

General Texts

1. **Austerweil, G. V.,** *Ion Exchange and the Exchangers,* Gauthiers-Villars, Paris, 1955.
2. **Dorfner, K.,** *Ion Exchangers, Properties and Applications,* Ann Arbor Science, Ann Arbor, 1972.
3. **Helfferich, F.** *Ion Exchange,* McGraw Hill, New York, 1962.
4. **Kunin, R.** *Ion Exchange Resins,* Robert E. Krieger Pub. Co., Huntington, N.Y., 1973.
5. **Nachod, F. C. and Schubert, J., Eds.,** *Ion Exchange Technology,* Academic Press, New York, 1957.

Specialized Books

A. Analytical

1. **Inczedy, J.,** *Analytical Applications of Ion Exchangers,* Pergamon Press, New York, 1966.
2. **Khym, J. X.,** *Analytical Ion Exchange Procedures in Chemistry and Biology,* Prentice Hall, Englewood Cliffs, N. J., 1974.
3. **Samuelson, O.,** *Ion Exchange Separations in Analytical Chemistry,* John Wiley & Sons, New York, 1963.
4. **Walton, H. F., Ed.,** *Ion Exchange Chromatography,* Dowden, Hutchinson & Ross, Stroudsburg, Pa., 1976.

B. Educational and School Texts

1. **Korolkov, N. M. and Vitols, O.,** *Synthesis of Ion Exchangers — Text Book,* Riga, Latvia, U.S.S.R., 1968.
2. **Leikin, Yu A.,** *Textbook for the Course of Physico-Chemical Principles of Ion Exchangers,* Khim-Technol. Inst., Moscow, 1974.
3. **Salmon, J. E. and Hale, D. K.,** *Ion Exchange — A Laboratory Manual,* Academic Press, New York, 1959.
4. **Stamberg, J. and Radl, V.,** *Ion Exchange — Principles for Laboratory and Industrial Practice,* Statni. Nakl. Tech. Lit., Prague, 1962.

C. Special Ion Exchange Products

1. Activated Carbon

   Steenberg, B., *Adsorption and Exchange of Ions on Activated Charcoal,* Almquist and Wikselis, Uppsala, Sweden, 1944.

2. Cellulosic Materials

   Peterson, E. A., *Cellulosic Ion Exchangers,* Elsevier, New York, 1970.

3. Chelates

   Hering, R., *Chelate Forming Ion Exchangers,* Monograph on Ion Exchange, Vol. 3, Akad Verlag, Berlin, 1967.

4. Inorganic Ion Exchangers

   Amphlett, C. B., *Inorganic Ion Exchangers,* Elsevier, Amsterdam, 1964.

   Nikolskii, B. P., Ed., *Inorganic Ion-Exchange Materials,* Izd. Leningrad University, 1974.

5. Membranes

   Laskorin, B. N., Smirnova, N. M., and Gantman, M. N., *Ion Exchange Membranes and Their Use,* At. Nauk. Tekh., Moscow, 1961.

   Saldadze, K. M., Ed., *Ion Exchange Membranes for Electrodialysis,* Khimia, Leningrad, 1970.

   Selegny, E., Ed., *Charged Gels and Membranes,* D. Reidl, Dordrech, Holland, 1977.

6. Redox Resins

   Cassidy, H. G. and Kun, K. A., *Oxidation-Reduction Polymers,* Interscience, New York, 1965.

   Kozhnikov, A. V., *Electron Ion Exchangers,* Khimiya, Leningrad, 1972.

D. Applications

1. Biochemistry and Organic Chemistry

Calmon, C. and Kressman, T. R. E., Eds., *Ion Exchangers in Organic and Biochemistry,* University Microfilm, Ann Arbor, Michigan 1957.

Samsonov, G. M., Trostyanskaya, E. G., and Elikin, G. E., *Ion Exchange Sorption of Organic Materials,* Moscow, 1971.

2. Catalysis

Isagulyants, V. I., *Ion Exchange Catalysis,* Izd. Akad. Naur. Arm., Yervan, U.S.S.R., 1975.

Polyansikii, N. G. and Sapozhnikov, V. K., *Ion Exchangers and Catalysis,* Znanie, Moscow, 1976.

3. Food and Beverages

Gerasimenko, A. A., Abramova, M. A., and Galovin, P. V., *Ion Exchange Resins in the Food Industry,* Izd. Akad Nauk Unv., Kiev, 1962.

Ionescu, T. D., *Ion Exchange in Chemistry and Food Technology,* Editura Tech., Bucharest.

O'Shanova, K. M., *Utilization of Ion Exchangers in the Food Industry,* Latvian Tekh. Inform., Riga, 1961.

Smit, P., *Ion Exchangers and Absorbers in the Manufacture and Refining of Sugars, Pectins and Related Materials,* Monograph on Ion Exchange, Vol. 5, Akad. Verlag, Berlin, 1969.

4. Medical

Martin, G. J., *Ion Exchange and Adsorption Agents in Medicine,* Little, Brown, Boston, 1955.
Miner, R. W., Ed., *Ion Exchange Resins in Medicine and Biological Research,* New York Academy of Sciences, 1953.

5. Metallurgical

Klimenko, N. G., Ivanovskaya, V. P., and Kalashmikova, T. M., *Use of Ion Exchangers to Improve the Selectivity of Flotation Process,* Nedra, Moscow, 1974.

Korkisch, J., *Modern Methods for the Separation of Rarer Metal Ions,* Pergamon Publishing, Elmsford, N.Y., 1969.

Tseft, A. L., Ed., *Ion Exchange Technology in Hydrometallurgy,* Nauk. Alma Ata., 1970.

6. Nuclear

Egorov, E. V. and Makarova, S. B., *Ion Exchange in Radiochemistry,* Atomizdat, Moscow, 1971.

International Atomic Energy Agency, *Operations and Control of Ion Exchange Processes for Treatment of Radioactive Wastes,* United Nations, 1967.

Lin, K. H., Use of Ion Exchange for the Treatment of Liquids in Nuclear Power Plants, Oak Ridge Natl. Lab., ORNL-4792-1973, 1973.

7. Soils and Minerals

Chelischev, N. F., *Ion Exchange Properties of Minerals,* Nauk, Moscow, 1973.

Kelley, W. P., *Cation Exchange in Soils,* Reinhold, New York, 1948.

8. Solvent Extraction and Liquid Ion Exchangers

General Mills Chem. Co., Feasibility of Liquid Ion Exchangers for Extracting Phosphate from Waste Water, U.S. Government Printing Office, Washington, D.C., 1970.

Marcus, Y. and Kertes, A. S., *Ion Exchange and Solvent Extraction,* Wiley-Interscience, New York, 1969.

Marinsky, J. A. and Marcus, Y., *Ion Exchange and Solvent Extraction,* Marcel Dekker, New York, 1973.

9. Water
Applebaum, S. B., *Demineralization by Ion Exchange* Academic Press, New York, 1968.

Arden, T. V., *Water Purification by Ion Exchange,* Plenum Press, New York, 1968.

E. Theoretical

1. **Issleib, K.**, Ed., *Proc. of Symposium on Anomalies in Ion Exchange Process,* Akad. Verlag, Berlin, 1962.
2. **Kakotov, Yu and Pasechnik, V. A.**, *Ion Exchange Equilibria and Kinetics,* Khimiya, Leningrad.
3. **Korolkov, N. M.**, *Theoretical Principles of Ion Exchange Technology,* Liesma, Riga, 1968.
4. **Marcus, Y. and Bowery, D. G.**, *Ion Exchange Equilibrium Constants,* I.U.P.A.C., New York, 1975.
5. **Soldatov, V. S.**, *Simple Ion Exchange Equilibrium,* Nauk. Teckh., Minsk, 1972.

Literature Reviews

1. **Kunin, R. et al.**, Ion exchange unit operations reviews, *Ind. Eng. Chem.*, 1948 to 1968; Ion exchange, *Anal. Chem.*, 1949 to 1966.
2. **Walton, H. F.**, Ion exchange, *Anal. Chem.*, 1970, 1974, 1976, 1978.
3. Ion Exchange Survey, Zerolit Ltd., for 12 years to 1969.
4. Russian Chemical Reviews, Reviews of Chelates, Pharmaceutical and Catalytic Applications.
5. **Boyd, G.**, *Annu. Rev. Phys. Chem.*, 1951.
6. **Bauman, W. L., Andersen, R. E., and Wheaton, R. M.**, *Annu. Rev. Phys. Chem.*, 1952.
7. **Juda, W., Marinsky, D., and Rosenberg, N. W.**, *Annu. Rev. Phys. Chem.*, 1953.
8. **Jandera, P. and Churacek, J.**, *J. Chromatogr.*, 86, 351, 1973; 86, 423, 1973; 98, 1, 1974; 98, 55, 1974.

Commercial Books

1. **Bayer, A. G.**, The Lewatits.
2. Dow Chemical Company, Dowex Ion Exchangers (Testing Procedures).
3. Diamond Shamrock, Duolite Ion Exchange Manual.
4. **Rohm and Haas**, *Ion Exchange Resins,* Amberlite Engineering Manual.
5. British Drug House Ltd., (Poole, England), Ion Exchange Resins.

Symposia

1. Ion exchange and its application, *Soc. Chem. Ind. (London),* 1955.
2. Ion exchange in the process industries, *Soc. Chem. Ind. (London),* 1969.
3. The theory and practice of ion exchange, Soc. Chem. Ind. (London), 1976.
4. Gordon Research Conference, since 1949, not published.

Nomenclature and Test Procedures

1. Recommendation on Ion Exchange Nomenclature, IUPAC, *Pure Appl. Chem.*, 1972.
2. American Society Testing Materials, Philadelphia.

Journals

1. Ion Exchange and Membranes, Gordon S. Breach (only two volumes)
2. Desalination and Membranes, Elsevier, New York.

APPENDIX B

ION EXCHANGER AND EQUIPMENT MANUFACTURERS

Commercial grades	Nomenclature of exchangers
United States	
Diamond Shamrock, Functional Polymer Division[a]	Duolite®
Dow Chemical Co.	Dowex®
Ionac Chemical Co., Division of Sybron	Ionac®
Rohm and Haas Co.[a]	Amberlite®
Europe	
France	
Dia Prosim (Ion Exchange taken over by Diamond Shamrock)	Allassion® Duolite®
Germany, West	
Farbenfabriken Bayer A. G.	Lewatit®
Permutit A. G.	Permutit®
Great Britain	
Jos Crosfield	Resex®; Resanex®
Zerolit Ltd. (Division of Diamond Shamrock)	Zerolit®, Zeo Karb®; De Acidite®
India	
Tulsi Industries	Tulsion®
Israel	
Ayalon	Srafion®
Italy	
Montecatini	Kastle®
Resin D'Ion (Division of Sybron)	Relite®

[a] manufacturer of exchangers in other countries also

Japan

Japan Organo (distributor of Rohm and Haas resins)	Amberlite®
Mitsubishi Chem. Ind. Ltd.	Diaion®
Nippon Soda	Nisso Alm®
Unitica Ltd.	Unicellex®

Netherlands

Imacti (Akzo Chem.)	IMAC®; Dusarit®; Asmit®

Government-Owned Companies (Europe)

Czechoslovakia (Chemapol)	Ostion®
East Germany (VEB Farbenfabriken Wolfen)	Wofatit®
Hungary (Chemolimpex)	Varion®
Poland	ESCARBO®
Russia	Letters (e.g., KV, AV)

MANUFACTURERS OF ION-EXCHANGE MEMBRANES

United States

Ionac Chem. Co., Birmingham, N.J. 08011
Ionics, Watertown, Mass. 02172

Japan

Asahi Glass
Asahi Chemical
Tokuyama Soda

MANUFACTURERS AND MAJOR DISTRIBUTORS OF LABORATORY AND SPECIAL GRADE ION EXCHANGERS

United States

J. T. Baker Chem.
Bio-Rad Laboratories
Brinkman International
Chemical Dynamics
DuPont de Nemours Co.
Durrum
Fisher Scientific
I.C.N. Pharmaceuticals
Mallincrodt
Pharmacia Fine Chemicals
Pierce Chemicals
Polysciences
Rohm & Hass

England

W & R Balston Ltd.

Germany

Macherey, Nagel Co. (cellulosic)
E. Merck A. G.
Schleicher and Schull (cellulosic)

MANUFACTURERS OF SPECIAL EXCHANGERS

Anion Exchanger Condensation Products
Bayer A. G.
Diamond Shamrock
Ionac
Rohm & Haas
Zerolit

Boron Removal Exchanger
Rohm & Haas

Carbamate Resins
Nippon Soda

Chelating Resins
Ayalon (Polyisothieuronuium Resin)
Bayer A. G.
Diamond Shamrock
Dow Chemical
Nippon Soda (Carbamate Resins)
Rohm & Haas
Unitica Ltd.

Cloth
Carborundum

Colloid Removal Exchanger
Rohm & Haas

Dextrans
Pharmacia

Electron Exchanger
Chemolimpex (Hungary)

Extracting Resins
Bayer A. G.

Heavy Metal Salts
Bio-Rad

Modified Greensand
Hungerford and Terry

Pellicular
DuPont

Polymeric Adsorbents
Rohm & Haas

Pyridine Type
Ionac

Retardation Resin
Dow Chemical Co.

Synthetic Alumino-Silicates

amorphouse gels:
Culligan, Inc.
molecular sieves:
Linde Division of Union Carbide,
W. R. Grace and Co.
Norton (mordenite)

Thiol
Akzo Chemical Co
Diamond Shamrock

MANUFACTURERS OF LIQUID ION EXCHANGERS (UNITED STATES)

Archer Daniels Midland Co.
Dow Chemical Co.
Eastman Organic Chemicals
General Mills Inc.
Minnesota Mining & Manufacturing Co.
Monsanto Chemical Co.
Rohn & Haas
Sterwin Chemicals Inc.
Vanderbilt Co.
Victor Chemical Co.
Virginia Carolina Chemical Corp.
Union Carbide Chemical Co.

EQUIPMENT MANUFACTURERS (UNITED STATES)

Ajax International Corp., Santa Barbara, Calif. 93103
American Water Cond. Equipment, Conshocken, Pa. 19428
Aqua Chem Inc., Milwaukee, Wis. 53212
Barnstead Co., Boston, Mass. 02132
Belco Pollution Control Corp., Paterson, N.J. 07509
Calgon Corp., Pittsburgh, Pa. 15230
Chemical Separations Corp., Oak Ridge, Tenn. 37830
Clack Corp., Windsor, Wis. 53598
Cochrane, Division of Crane, King of Prussia, Pa. 19406
Culligan Inc., Northbrook, Ill. 60062
Eimco, Division of Envirotech, Salt Lake City, Utah 84110
Envirex, a Rexford Co., Waukesha, Wis. 53186
General Filter Co., Amex, Ind. 50016
Graver, Division of Ecodyne, Union City, N.J. 07083
Hungerford and Terry Inc., Clayton, N.J. 08312
Illinois Water Treatment Co., Rockford, Ill. 61105
Infilco-Degremont Inc., Richmond, Va. 23216
Ion Exchange Products Inc., Chicago, Ill. 60640
Ionics, Watertown, Mass. 02172
Los Angeles Water Treatment Co., Los Angeles, Calif.
Permutit, Division of Sybron, Paramus, N.J. 07652

Small Unit Manufacturers
- Barnstead, Boston, Mass. 02132
- Corning Glass Works, Corning, N.Y. 14830
- Culligan Inc., Northbrook, Ill. 60062
- Crystal Labs, Hartford, Conn.
- Millipore, Bedford, Mass. 01730
- Vaponics, Plymouth, Mass. 02360

APPENDIX C

SPECIAL TABLES OF DATA

Expression of Ionic Concentrations

U.S.: grain per U.S. gal, or parts per million (ppm)
United Kingdom: Clark, equivalent to a grain per British Imperial Gal
France: parts per 100,000 parts of water
Germany: part of CaO per 100,000 parts of water

	Equivalence of the above			
U.S.	English degrees (1 Clark)	French degrees	German degrees	parts per million
gr/U.S. gal	1.2	1.71	0.958	17.1
ppm	0.07	0.1	0.056	1.0
meq/ℓ	—	—	—	50 as $CaCO_3$

Calculations

Volume of exchanger required for a bed

$$\frac{\text{Volume of water to be treated between regenerations} \times \text{Concentrations of ions to be removed}}{\text{Rated capacity}}$$

Units to be used

If volume is to be in cubic ft, then volume of water is in gal. Concentration of ions in grains as $CaCO_3$/gal divided by 1000. Rated capacity in kgr as $CaCO_3$ per cubic ft.

If volume of resin is to be in cubic meters, then volume of water is in cubic meters. Concentration of ions and rated capacities in equivalents.

$$\frac{\text{Volume } (m^3) = \text{volume of solutions } (m^3) \times \text{Concentration of ions } (eq/\ell) \times 1000)}{\text{Rated capacity } (eq/\ell \times 1000)}$$

Rated capacity of an exchanger

$$\frac{\text{Volume of water treated} \times \text{Concentration of ions - leakage}}{\text{Volume of resin used}}$$

Regenerant Efficiency

Efficiency in the past was expressed in terms of pounds of regenerant per kilograin of capacity obtained. For example, if a resin gives 18 kgr per cu ft at 5 pounds of salt as the regenerant, the efficiency then is 5/18 = 0.28 lb. Theoretical value is 0.17 lb of NaCl per kgr as $CaCO_3$. Therefore, the actual compared to the theoretical is 0.28/0.17 = 1.64, or, 64% more salt was required than the theoretical. Today, the efficiency would be expressed as 1.64, i.e., 1.64 times the stoichiometric or theoretical value.

Numerical Values to be Noted

120 ppm = 1 lb per 1000 gallons
0.17 lb of salt = 1 kgr as $CaCO_3/ft^3$
1 kgr as $CaCO_3/ft^3$, = 45.8 meq/ℓ
To calculate requirement of theoretical regenerant dosage

$$\text{HCl: } \frac{45.8 \times 28.3}{454 \times 1000} \times 36.5 = 0.104 \text{ lb/kgr as } CaCO_3$$

$$H_2SO_4\text{: } 0.002855 \times 49 = 0.140 \text{ lb/kgr as } CaCO_3$$

$$\text{NaOH: } 0.002855 \times 40 = 0.114 \text{ lb/kgr as } CaCO_3$$

$$\text{NaCl: } 0.002855 \times 58.5 = 0.167 \text{ lb/kgr as } CaCO_3$$

STANDARD SIEVE OPENINGS FOR VARIOUS MESH NUMBERS

Mesh number	U.S. sieve opening in mm	British sieve opening in mm
10	2.00	1.50
16	1.19	—
20	0.84	0.735
25	0.71	—
30	0.59	0.485
40	0.42	0.360
50	0.30	0.286
60	0.25	0.235
80	0.18	0.173
100	0.15	0.136
200	0.074	0.061

Definitions

Effective size — The screen size (i.e., millimeter opening of screen) which passes 10% of the representative sample, or at which 90% of the sample is retained.

Uniformity coefficient — The ratio obtained by dividing the sieve opening in millimeters at which 60% of the sample passes by the mesh size of the screen (also in millimeters) which passes 10% of the sample.

Conversion of Units

Linear		
in.	2.540 cm	0.3937
ft	30.48 cm	0.0328
yd	91.44 cm	0.01094
mi	1.609 km	0.6214
μm	1×10^{-4} cm	10,000
Area		
in^2	6.452 cm^2	0.1550
ft^2	929 cm^2	1.08×10^{-3}
ft^2	0.0929 m^2	10.76
$mile^2$	2.590 km^2	0.3861
acre	4047 m^2	2.47×10^{-4}
acre	0.001562 mi^2	640
Volume		
in^3	16.39 cm^3	0.0610
ft^3	0.0283 m^3	35.31
ft^3	28.317 ℓ	0.0353
ft^3	7.47 gal (U.S.)	0.134
gal (U.S.)	3.785 ℓ	0.2642
gal (Imperial)	4.545 ℓ	0.2200
acre-foot	1233.6 m^3	8.11×10^{-4}
Weight		
gr (t)	0.0648 g	15.43
oz (a)	28.349 g	0.0353
lb	453.6 g	2.205×10^{-3}
ton	907.18 kg	1.1023×10^{-3}
ton	0.9842 ton (metric)	1.0161
Concentration		
gr/gal (U.S.)	17.118 ppm	.0585
gr/gal (U.S.)	0.017118 g/ℓ	58.417
lb/gal (U.S.)	119.8 kg/m^3	0.00834
lb/ft^3	16.026 g/ℓ	0.0624
ppm	0.001 g/ℓ	1000
mg $CaCo_3$/liter	0.02 meq/ℓ	50
Flow Quantities		
ft^3/min	472 cm^3/sec	0.00212
ft^3/min	1.6996 m^3/hr	0.5883
ft^3/min	7.4806 gal (U.S.)/min	0.13368
ft^3/min	0.1247 gal (U.S.)/sec	8.025

ft/min	0.508 cm/sec	1.98
ft/min	30.48 cm/min	0.3281
gal (U.S.)/min	0.06309 ℓ/sec	15.85
gal (U.S.)/min	0.2271 m^3/hr	4.403
gal (U.S.)/ft^3	133.74 ℓ/m^3	0.00748
gal (U.S.)/ft^3	0.134 ℓ/ℓ	7.48
gal (U.S.)/min/ ft^2	2.45 m^3/hr/m^2	0.408
gal (U.S.)/min/ ft^3	8.02 m^3/hr/m^3	0.125
lb/ft	14.88 g/cm	0.0672
lb/ft^2	4.887 kg/m^2	0.2048

Note: To convert unit in second column to units in first column, multiply by value in the third column.

INDEX

A

B

C

D

E

F

G

H

I

J

K

L

M

N

O

P

Q

R

S

V

W

X

Z